SORRY FOR THE INCONVENIENCE
BUT THIS IS AN EMERGENCY

LYNNE JONES

Sorry for the Inconvenience But This Is an Emergency

The Nonviolent Struggle for Our Planet's Future

HURST & COMPANY, LONDON

First published in the United Kingdom in 2024 by
C. Hurst & Co. (Publishers) Ltd.,
New Wing, Somerset House, Strand, London WC2R 1LA
Copyright © Lynne Jones, 2024
All rights reserved.

The right of Lynne Jones to be identified as the author of
this publication is asserted by her in accordance with the
Copyright, Designs and Patents Act, 1988.

Distributed in the United States, Canada and Latin America by
Oxford University Press, 198 Madison Avenue, New York, NY 10016,
United States of America.

A Cataloguing-in-Publication data record for this book
is available from the British Library.

ISBN: 9781911723035

This book is printed using paper from registered sustainable
and managed sources.

www.hurstpublishers.com

'Letter to a Woman on the Falklands Victory Parade' (1982) first appeared in *Greenham Common: Women at the Wire* (London: Women's Press); 'Breaking Barriers' (1983), 'In the Eye of the Storm' (1983), 'Shut Up and Listen' (1984) and 'Changing Ideas of Authority' (1984) first appeared in *the New Statesman*; and 'The Woman Behind Solidarity' (1984) first appeared in *Ms Magazine*.

Printed in Great Britain by Bell and Bain Ltd, Glasgow

Disobedience, in the eyes of anyone who has read history, is man's original virtue. It is through disobedience that progress has been made, through disobedience and through rebellion.

Oscar Wilde, 'The Soul of Man Under Socialism'

For Asmamaw

CONTENTS

LIST OF ABBREVIATIONS

A & E   Accident and Emergency
AWRE   Atomic Weapons Research Establishment
CND   Campaign for Nuclear Disarmament
DFID   Department for International Development
ECHR   European Convention on Human Rights
ECtHR   European Court of Human Rights
GAM   Grupo de Apoyo Mutuo
ILO   International Labour Organization
IPCC   Intergovernmental Panel on Climate Change
JSO   Just Stop Oil
PBI   Peace Brigades International
XR   Extinction Rebellion

ACKNOWLEDGEMENTS

First of all, I want to thank all the nonviolent activists, past and present, who have inspired me to write this book. I have accompanied or witnessed some of them on the actions documented here; many agreed to be interviewed. I hope this book does justice to all your work and commitment. I am also grateful to all the writers and journalists who have documented and written about nonviolent activism in different countries, at different times and in different forms. Your work has contributed greatly to my own.

Patrick Burke, Amelia Cussans, Stuart Drysdale, Kathy Fallon, Chris Newman and Richard Vogler all read and commented on sections or all of this book, thank you. Thanks also to Ali Rowe for keeping me updated with new developments and new source materials.

Thanks to Silvia Weber, Carlos Juárez, and Bernardo and Maria Caal Xol for facilitating work in Guatemala and agreeing to be interviewed. Thanks also to everyone in the Peaceful Resistance of Cahabón for your hospitality and making my husband and myself so welcome.

I am very grateful to the *New Statesman*, first for commissioning me to keep a regular diary about the Greenham Common Women's Peace Camp during the 1980s; and secondly, for allow-

ACKNOWLEDGEMENTS

ing me to republish some of those diaries here. Thank you to Michael Randle and Irene Publishing for permission to republish parts of Michael Randle's speech at the Old Bailey from *Rebel Verdict*. I would like to thank *Ms Magazine* for permission to republish my 1984 interview with Anna Walentynowicz in Poland. I am also grateful to the Russell Press and the Bertram Russell Peace Foundation for permission to use the quote from E. P. Thompson's *Protest and Survive* as an epigraph.

Many thanks to Lara Weisweiller-Wu for commissioning this book, and to Alice Clarke for supportive and skilful editing. Thanks also to Rose Bell and Jon Idle for such scrupulous copy editing and proofing respectively; and to all at Hurst for bringing it to publication.

Thanks very much to the Morrab Library in Penzance and all the staff and volunteers for providing such a beautiful, tranquil home in which to write.

Finally, love and thanks to my husband Asmamaw Yigeremu, who shares my concerns about the issues discussed, and has engaged in many of the actions with me. His interest, reflections and feedback on my writing throughout, and his constant support and accompaniment on this journey, are what has made it possible.

INTRODUCTION

The planet burns, people are drowning. The ice caps melt and the forests that once cooled the air now add to its warmth. On every news channel apocalyptic scenes of fire, flood, drought and devastation bring home a grim reality that is of our own making. Meanwhile, on YouTube, I replay a pale schoolgirl from Sweden addressing the great and the good at the United Nations in September 2019:

> You have stolen my dreams and my childhood with your empty words. And yet I'm one of the lucky ones. People are suffering. People are dying. Entire ecosystems are collapsing. We are in the beginning of a mass extinction, and all you can talk about is money and fairy tales of eternal economic growth. How dare you![1]

She is right. We have messed up. And I too am afraid. The last time I felt this frightened was in the early 1980s. The Soviet Union had invaded Afghanistan, martial law had been imposed in Poland, and new nuclear weapons were about to be deployed across Western Europe with the express purpose of fighting a nuclear war. Riots in Toxteth and Brixton had demonstrated the deep divisions, inequality and racism at home. It was impossible to avoid a sense of impending global catastrophe.

I was terrified. I had to do something. In 1982 I resigned my job as a young casualty officer in a Liverpool hospital and went

1

to live in a women-only encampment outside an airbase at Greenham Common, to protest the proposed siting and deployment of those new nuclear weapons. For the next decade I was fully engaged in political activism across Europe and in Central America as part of the nonviolent social movements that contested what historian E. P. Thompson called the exterminating logic of the Cold War.

This book describes why and how I became engaged in nonviolent civil disobedience as a means of political change, and explores its use in the nonviolent protest movements of today. The book is not a manual, it is not a call to action, it is not a 'how to do it' book. It is one woman's personal reflections on what are the best ways to act in a Western democracy, given what we face. In thinking that through I draw on and describe past experiences of dealing with existential threats, and then share my own journey over the last five years exploring the use of nonviolent protest in the face of the current situation, where the existential threat of nuclear weapons has not gone away but is combined with the climate and ecological emergencies, and the global systems of power and injustice that drive them.

Forty years ago, a small bunch of women marched from Cardiff to a proposed nuclear missile base in Berkshire and sat down to stay because they were ignored. They inspired and empowered thousands of other women to join them and carry on the protest in spite of constant pressure to leave, until those missiles were removed. That story may be forgotten, but the forms of resistance chosen: site occupation and continuing nonviolent confrontation, were copied around the world and continue to be used. Today, in the face of the looming global catastrophe caused by the climate and ecological crises, the mass extinction of animal species and the rise of fascism around the globe, there is a new surge of nonviolent direct action and policies are changing as a result.

INTRODUCTION

Why tell these stories now? Do they matter?

Yes. First, because social movements in the 1980s showed that nonviolent direct action works. By the end of the decade not only had those new nuclear weapons been removed, but the 'Iron Curtain' was torn down, and the division of Europe ended. Nonviolent resistance can transform a continent. Second, because we failed to sustain those victories. The weapons have been recycled and endure. The treaties are torn up. Authoritarian governments re-emerge. Disasters and conflict now forcibly displace one person every two seconds.[2] There is no 'end of history'. The Doomsday Clock set by the Bulletin of the Atomic Scientists stood at four minutes to midnight in the early 1980s. In 2022 it was set at 100 seconds to midnight because of the combined threats of accelerating nuclear weapons programmes, the climate crisis, government failures to tackle these problems, and the growing impacts of disinformation and loss of trust in science. In 2023, because of the added threat from the war in Ukraine, the clock now stands at 90 seconds to midnight, the closest it has ever been.[3]

So the methods we used—active, assertive, creative nonviolent direct action—are now more necessary than ever. Action is urgent, but we also need to think, to reflect on history and discuss which methods are best and most effective. These will vary according to time, culture and context, and it is unlikely that one size fits all. How can we learn from past experience and also from each other?

Third, the discussions that Greenham women had in the 1980s about the connections between gender, power and violence have never been more pertinent. The #MeToo movement began to draw back the curtain on the misogyny and profound abuse of power that exists across society, while Black Lives Matter has exposed and confronted racism. Both have to be addressed. Greenham women pointed out that patriarchy oppresses both men and women by confining them, at home and at work, in gender-defined straitjackets, constricting and distorting the thoughts, feelings and behaviours of all of us. This leads directly to the situation we have today: a world out of control where authoritarian rulers prevail, East and West, North and South. We all need to be feminists now.

At a time when democracy appears to be under attack from all sides and the future of life itself is at stake, stories of successful resistance are needed. History is an antidote to despair. What Greenham taught me was that we are all much more powerful than we realise. The smallest acts of nonviolent resistance by one individual can make a difference. In describing and reflecting on the activism of that decade, by drawing on other historic examples to explore their relevance to the crises of today, and by giving voice to present day activists, both in Britain and elsewhere, I want to encourage both more nonviolent action and hope. At Greenham we argued that every single individual was of equal importance and that what she did mattered. It is still true today.

The book combines personal and public history, connecting past and present nonviolent activism. The main focus is on its use within a democracy, particularly the United Kingdom, but some examples from other countries are included. After an initial chapter in which I introduce myself and explain why and how I became involved in nonviolent action, each chapter is prefaced by a published article or personal diary from the 1980s, and then explores a particular theme. These include whether obstruction is an effective form of nonviolence; whether damage to property can be considered nonviolent; the role and use of the law in nonviolent struggles; the importance of direct nonviolent interventions; an exploration of different forms of non-cooperation; and what we can learn from nonviolent Indigenous environmental activists in the Global South. During the period of writing both actions and legislative responses have evolved and changed constantly, as has my own thinking about the best way to act. So I have allowed the chapters to document those changes as they occurred through time.

I should emphasise that the historical examples included are the ones that have inspired me. Many inspiring stories have been left out, as this is not a comprehensive history. I hope through storytelling to encourage others to find a way to take nonviolent action in the ways that work for them. Nor is this a definitive history of any of the political struggles detailed above. Thousands of us were involved at Greenham, and thousands more in the wider peace and human rights movements of that decade. The same is true with the social movements in which I am engaged today. Each has her or his own account. This is just one woman's story, my own. It is a truthful account. Reality matters in these days when every fact is contested.

1

ACCIDENTS AND EMERGENCIES

Two Conservative MPs have calculated that effective measures might reduce deaths in a nuclear war in this country from about thirty-five millions to twenty millions. [...] There will be bunkers deep under the Chilterns for senior politicians, civil servants and military, and deep hidey holes for regional centres of military government. That is very comforting. The population of this country, however, will not be invited to these bunkers, and it is an Official Secret to say where they are.

E. P. Thompson, *Protest and Survive*, 1980[1]

The phone rang in the doctors' mess. Dr P. put down the pool cue and went to pick it up.

'Red eye? Female? Ok, how old? Seventeen?' He smiled. 'Ok, I'll be there right away.'

'What is it?' said J. eyeing the balls on the table.

'Red eye, hey what's that condition where you have red eye and ulcers on your fanny?'

'Behçets? It's very rare. Ulcers all over.'

'Behçets, exactly, better examine her fanny just in case.' Dr P. straightened his tie, pushed back his slightly too long blonde hair. 'I'll wait a minute, better not to look too eager.'

I bit my lip, kept my head down and scrawled a note to add to my collection, determined not to react. Yesterday, J. had been complaining about a young woman wheeled in with a hip injury after being hit by a car. Apparently, she was not attractive enough. He missed his old job where his boss had been in charge of all the breast cases for the unit: '50 per cent of the admissions were young girls for minor routine ops, perfectly healthy, but you had to listen to their chests, amazing!'

We were all junior doctors in the Accident & Emergency department of a large northern hospital. An unusual lull in admissions had allowed us to retreat to the mess. Curled up in an orange vinyl armchair by the hissing gas fire, I had enough to think about. I was meant to be writing a speech for a public meeting the following week. My job was to explain to any medical staff who cared to attend what blinding flashes, blast, heat and radiation from a nuclear weapon could do to soft human bodies. Hopefully I could terrify them into joining the newly formed Medical Campaign against Nuclear Weapons.

It was E. P. Thompson's fault. A year previously, in 1981, standing at W. H. Smith at Kings Cross railway station in London, I had bought a pamphlet to while away a train journey and instead found the world closing in. According to Thompson, in December 1979 my government had made a decision, in secret, to put new nuclear weapons in Britain. They were called cruise missiles, and the idea was that if the West appeared to be losing a conventional war, they would be used. Moreover, these missiles would not be under our control. They would be 'owned and operated by the United States'. Britain was about to become an expendable aircraft carrier, available at any moment for a friendly little 'theatre nuclear war' that could be 'limited' to Europe.[2]

As Thompson explained, our survival would apparently depend on reading a little government-issued pamphlet called *Protect and Survive*, which explained how to build a fallout shelter under the stairs using old planks, doors and heavy furniture 'then creep into these holes with food and water for 14 days, a portable radio, a portable latrine, and of course, a tin opener'. Unfortunately this would be unlikely to protect very many of us from the intense heat, blast and radioactive emissions of an exploding nuclear weapon: 'We must envisage many thousands of nuclear families listening to Mr Robin Day's consensual homilies on their portable radios as they are burned, crushed or suffocated to death.'[3]

Apparently, the deaths of millions were regarded as acceptable in this new war-fighting strategy. Nor was it confined to Britain. Similar theatre nuclear weapons would be deployed across Europe. Thompson had 'come to the view that a general nuclear war is not only possible but probable, and that its probability is increasing'. His anger bounced off the page and sparked a reciprocal rage in me. I felt furious at both my ignorance and the feeling that I had been asleep. How could it be that I knew nothing about this? Could it be stopped?

My parents had taken me on the early Ban the Bomb marches in a pushchair, and I could remember as a small child, during the Cuban missile crisis, praying that the atom bomb would not fall that night. But since then, the fact that both sides in the Cold War had sufficient weapons to wipe out civilised life on the planet many times over, and yet still continued to develop new and more deadly bombs, had passed me by. Then the Soviets invaded Afghanistan in December 1979. I was skiing with friends in January, and heard someone in the queue for the ski lift joking that it was a good thing we were in Switzerland because US President Jimmy Carter had delivered an ultimatum to the Russians, saying that he had initiated military call-up and if they made one further move, he would declare war.[4] We spent the evening trying to get more information from the BBC World Service and giggling about the possibility of a nuclear exchange. There were rumours that Prime Minister Margaret Thatcher had responded to that speech by putting the whole country on the highest level of nuclear alert, meaning that those bunkers from which survivors would be governed in the post-apocalyptic world were actually being made ready.

Suddenly nuclear war was the topic of the moment. There were debates in parliament and letters to newspapers. Eminent men spoke out: Lord Louis Mountbatten, not known for his pacifism, had made a speech to the Stockholm Institute for Peace Research shortly before he was assassinated by the IRA, in which he declared that he did not believe

> That any class of nuclear weapons can be categorised in terms of their tactical or strategic purposes. [...] In the event of a nuclear war there will be no chances, there will be no survivors—all will be obliterated. I am not asserting this without having deeply thought about the matter. When I was Chief of the British Defence Staff, I made my views known [...] I repeat in all sincerity as a military man I can see no use for any nuclear weapons which would not end in escalation, with consequences that no one can conceive.[5]

I wasn't the only one who felt alarmed. Local peace groups sprang up all over the country. When E. P. Thompson came to speak in the small cathedral town where I was working in the spring of 1980, it was standing room only. He was an extraordinary man: a Marxist who had left the Communist Party in disgust over the Soviet invasion of Hungary in 1956, and a historian who had completely revolutionised the study and understanding of history with the publication of *The Making of the English Working Class* in 1963. Even my history teacher, in my private school in the late 1960s, encouraged us to read it. She wanted us to understand that the struggles of weavers and farmers and factory workers for a better life mattered as much as the lives of kings and queens and the wars they fought. Now Thompson, like a latter-day John Wesley, was barnstorming the country, galvanising a whole new resistance movement into being.

ACCIDENTS AND EMERGENCIES

I listened from the crowded gallery as this tall, eloquent man, with angular features and a shock of white hair, made the same simple, persuasive points he had made in his pamphlet. My generation, growing up in the shadow of the Cold War, had become habituated to the idea of nuclear weapons as part of the status quo. But we needed to recognise that the policy of deterrence: mutually assured destruction (MAD), was not stable but a 'degenerative state', and that actual nuclear war, whether by accident or design, was increasingly likely in our lifetimes. We had to ask and answer difficult questions: first, was the deaths of millions preferable to occupation which might offer the possibility of resistance and recovery? Second, could we endorse the use of such weapons against innocent people? And third, how come we were committed to policies that endangered the survival of the nation without public or parliamentary discussion?[6]

The first two questions were easy to answer: No and No! It was the third question that enraged me. That decision, taken on 12 December 1979, brought home in stark relief the fact that we had absolutely no knowledge about, and no say or control over, life and death decisions that affected us all.

Thompson urged revolt, and not only against the deployment of new nuclear weapons—that was just a starting point. It should be against the antagonism and military structures that justified them. He urged us to sign the 'Appeal for European Nuclear Disarmament' that he and others at the Bertrand Russell Peace Foundation had written:

> We must commence to act as if a united, neutral and pacific Europe already exists [...]. We must learn to be loyal, not to 'East' or 'West' but to each other, and we must disregard the prohibitions and limitations imposed by any national state. [...] Our objectives must be to free Europe from confrontation, to enforce detente between the United States and the Soviet Union, and, ultimately, to dissolve both great power alliances.[7]

Thompson wanted 'Détente from below'. Unilateral disarmament initiatives were part of that. Local groups across Europe could take on their own national governments in their own way, but also reach out and collaborate with each other. Together we would end the Cold War.

I have never been a joiner. I tend to run in the opposite direction when offered the chance to belong to a group. When asked to define my beliefs in my student years, I always found it easier to say what I was not. The 'God Squad', whipping out their bibles over tea and crumpets in my first year at Lady Margaret Hall, Oxford, and asking if I had been 'saved', quickly decided I was beyond redemption when I told them I attended Quaker meetings. I avoided student politics; neither authoritarian Socialist Workers or foppish Young Conservatives appealed. Announcing I was an anarchist and a pacifist produced looks of incredulity and did not require group membership of any description. I preferred concrete campaigns on the issues that upset me. I campaigned against Apartheid, because at medical school in Bristol I lived with a Black South African refugee medical student who taught me what it meant to experience its horrors; but further political engagement consisted of reading Emma Goldman, Peter Kropotkin and Mikhail Bakunin and helping to staff an anarchist bookshop.

Thompson made the threat of nuclear war seem imminent and personal, and his call for a Nuclear Free Europe, connecting peace and human rights, caught my imagination. I signed the Appeal, joined my local Campaign for Nuclear Disarmament (CND) group, started leafleting in the marketplace at weekends and went on marches. But I wanted to do more, it felt too urgent to wait, and fortunately others felt the same way. Howard Clark was living in the area. A lifelong pacifist and a former editor of the pacifist magazine *Peace News*, he had the glamour of having actually faced arrest through his involvement in the campaign to withdraw British troops from Northern Ireland. In 1978 he had organised nonviolence training for some 4,000 people who occupied the site of a proposed nuclear power station at Torness in Scotland. Now he suggested setting up a nonviolent direct action group for the peace activists in the city.

So as spring moved into summer, some of us formed an 'affinity group', the term Howard used for a small group who could take actions together because they knew and trusted one another and shared common goals. We met weekly to train in civil disobedience. We discussed when and where nonviolent law-breaking was justified, and what sort of actions we would carry out: blockades, sit-ins and other forms of non-cooperation. We learnt how important it was to know one's legal rights, and to prepare for arrest in advance. We role-played being manhandled, arrested and searched by the police, de-escalating conflict in difficult situations, and group decision-making in stressful ones. We took turns in facilitating meetings, and I learnt that essential skill of a good facilitator: have no investment in a particular outcome from any meeting you facilitate; your task is to help others reach a decision by giving everyone space to speak, but to keep your own opinions to yourself.

Role-playing nonviolent protests in the evenings provided an antidote to my day job. When medical students finished their studies, they were expected to complete a year as a pre-registration junior house officer. (These days the military terminology has been dropped and they are called F1s.) This requires six months in a general medical post and six months in a surgical one. I was attached to the surgical unit of a busy district general hospital. My mornings were mostly spent clerking in patients and ensuring they were fit for theatre, taking the bloods, and arranging any laboratory tests and X-rays that were needed. Afternoons were spent holding retractors and helping to stitch-up in theatre; evenings on call checking up on post-operative care and admitting emergencies. I was on call every third night and weekend, which meant that by 5 p.m. on Monday I might have gone eighty hours without sleep. I was working for a skilled and competent boss who prided himself on the shortness of his waiting list, which, on day-case days, could mean seeing eight patients in two hours. Many of them appreciated the efficiency; they just wanted their varicose veins fixed. But the predominant attitude to elderly patients was typified by a conversation I overheard at lunch one day between two young male junior surgeons:

'Poor John didn't get a break all night, one bloody patient to the next.'

'Good character-building stuff though.'

'C. is sending in that 88-year-old.'

'Oh no!'

'Yeah, bloody gerrys.'

'Can't stand operating on gerrys myself.'

'The physicians are finally out of our beds.'

'Good!'

'You know what they did? C. rang me about that gerry. I refused to admit, so they rang the Responsible Officer, and he said they could have the bed! It's ok if they're dead in the morning. It's when they are bloody alive and still there.'

The hospital dining room was definitely not the place to discuss the politics of nuclear annihilation. By now, newspaper polling was telling us that 40 per cent of the British public thought nuclear war was inevitable, yet none of my colleagues shared my concerns. On the contrary, a nice, fresh-faced young houseman assured me over coffee that the nuclear destruction of civilisation and the loss of 25 million lives was definitely preferable to Russian occupation. At least it meant we were 'finally showing some muscle'.

I was beginning to understand how my colleagues in the God Squad might have felt about failing to save me. Here is the thing about a conversion experience, because that is what my engagement in the peace movement was like. You walk into an unknown room on a sunny day. Perhaps you are going to rent it. You still have your sunglasses on and it all looks fine. Then you take the sunglasses off: you suddenly notice things that had been invisible before. The ceiling is dipping and looks as if it might collapse, the walls are filthy, the curtains are ragged and torn, the window glass is broken, the floor is covered with cockroaches. You can put your sunglasses back on, but you cannot get rid of the knowledge that the room is dangerous, dirty and squalid. I could not undo my newly acquired knowledge of the nuclear threat and the risks it posed to us all, and I wanted everyone else to have the same clarity of vision. This makes for painful and boring company, unless you spend your time with the similarly converted. I could push it out of my mind for periods of time but I could not unknow it and this coloured everything.

Walks and bike rides in the astonishingly lovely moorland countryside outside the town on my weekends off would be filled with a poignancy and a certain bleakness. I could not avoid thinking: this can all be destroyed in an instant, by a small number of short-sighted men and women, out of fear. While clerking in children for operations on broken arms or legs, I would ask myself: and then what? What are we offering as a future? Sometimes I looked at atlases and fantasised about escaping to a safer place, New Zealand perhaps? But I had read Neville Shute's *On the Beach*; there was no escape, and anyway I felt responsible. If being a doctor was about caring for people's wellbeing, what could be more important than confronting nuclear annihilation?

The first opportunity for serious action came that August, after I had already finished my contract and gained my full registration. I had decided to take time off before applying for another medical job. Operation Square Leg was a 'home defence exercise' to see how prepared local authorities were for a Soviet nuclear attack. The scenario was that some 205 megatons would fall across Britain (think 10,000 Hiroshima bombs). This would immediately kill 29 million, severely injure 7 million, and around 19 million would survive, at least initially. The exercise would test how the planners would cope in their bunkers.[8]

The general public were not invited to participate, but my affinity group decided to join in. After all, what could be more likely than civilian survivors demanding refuge at the local county headquarters? Howard had somehow got hold of both the location and timing of the post-attack planning meeting and eight of us turned up and walked through the front doors. To our astonishment no one stopped us, so we proceeded down the stairs, at which point a flustered official appeared and begged us to retreat. We should have pressed on and burst into the planning room, hysterically demanding food, water and medical care, but we had not expected to get so far and had not prepared ourselves properly. Being nice, nonviolent types, we retreated back to the front door where we found the press waiting: not for us, but for Mr Leon Brittan, the Home Office minister who just happened to be visiting the exercise. Another lesson learnt: do not stand on the steps below a government minister and ask him questions when cameras are rolling. Remain on the upper step and make a speech.

That November I wrote a letter to *The Lancet* arguing that civil defence was a con to make the public think nuclear war was survivable and induce a false sense of security, thus allowing Britain to pursue an aggressive position that increased our vulnerability:

> Nuclear war involves a complete reversal of our code of practice. We would be expected to leave those in greatest distress and only attend to those with minor injuries [...] Doctors have a responsibility to educate themselves and others on the real dangers of thermonuclear war. We must point out that we are faced with the most insidious and terrible disease that has ever affected human beings. It is called the arms race. [...] In its final stages there is no cure, but prevention is still a possibility if we act now.[9]

The correspondence pages of various medical journals showed that I was not the only one thinking this way. A letter to the *British Medical Journal* at the end of that same month stated:

> Now that the deployment and use of 'tactical' nuclear weapons are proclaimed NATO policy, and civil defence against nuclear war is officially discussed and encouraged, the unthinkable has evidently become thinkable. Doctors of medicine and scientists in related fields know that nuclear weapons are so destructive to human life and health that they must never again be used. Prevention of nuclear war offers the only possibility for protecting people from its medical consequences.
>
> We have therefore initiated an organisation under the title Medical Campaign against Nuclear Weapons to heighten the awareness of the medical profession and those concerned with public policy about the medical implications of nuclear war.[10]

I joined the campaign as soon as I came home from travelling in the spring of 1981. I immersed myself in learning and understanding those devastating medical effects: the distance at which retinal burns occurred, precisely how much radiation killed you quickly, or slowly. Soon I was sharing this knowledge at least once a week at public meetings around the country. CND membership was increasing exponentially and it felt like everyone I knew was forming or joining something with 'anti-nuclear' in the name. My cellist friend was playing for Musicians against Nuclear Arms. My lawyer friends were busy discussing whether nuclear weapons were illegal.

The trouble was that standing up and telling people what happens to their eyeballs when they stare at the flash from the exploding bombs is emotionally draining and exhausting. I could see the benefits, but I was increasingly worried that trying to terrify people into action was the wrong approach. Fear was not enough. 'I have the sensation that I am standing stock still in the middle of a hurricane weeping,' I wrote to a friend: 'How come I think I can succeed when civilisation as a whole has brought us to this point? Working for peace should surely be positive not negative, a small statement about one's hopes for the world, but instead I find my life is an anguished cry of rage, hoping I will be heard before it is too late.'

Meanwhile, I had just obtained a job as a senior house officer in an A & E department: six months with adults, six months with children. I was one of two women appointed alongside two men. Then, just a few weeks before I started work, a woman called Ann Pettitt rang me up from Wales. She was organising a Women's March, to walk from Cardiff to the United States Air Force base at Greenham Common in Berkshire, which was now one of the designated sites for cruise missiles. She had read an article I wrote for *Peace News* about an action in which I had participated in the United States the previous November, where some 2,000 women had completely encircled and blockaded the Pentagon. Would I like to come on the march? They were calling themselves 'Women for Life on Earth'. When I explained that I could not take the time off from my new job, Ann invited me to meet them in Bristol when I was off duty to give a talk.

Standing in the pedestrian area next to the shopping centre in Bristol, with traffic roaring by and tired Saturday shoppers paying little attention, I made my speech to some forty women and children and a handful of men. I pulled together all the feelings and thoughts that had been bubbling away for the previous two years. Parts still seem as relevant today:

> We live in a society whose organising principles are violence and a disregard for human life. That is the real problem and it seems to me that by calling yourselves Women for Life on Earth you see that, and are beginning to come close to a solution. I have a name for such a society—it is called a patriarchy [...] when women talk of Liberation, it's not from men, but from a patriarchal value system that exploits us both and threatens to kill us both by denying us all half our humanity.

ACCIDENTS AND EMERGENCIES

> [We want] a society where the organising principle is nurturance and cooperation not violence and competition. [...] A society where human relationships have a greater value than material goods so that no one could espouse the neutron bomb as a useful weapon.

Then they marched on to the airbase, and I headed north to my A & E department.

I liked the work. The draining bureaucracy of the last job was removed. Instead of spending hours on the phone arranging surgical lists and chasing blood results, I spent eight hours daily seeing people in trouble: cut fingers and cut heads, monkey bites, broken arms, small children with polystyrene in the ears, heart attacks, wedding rings stuck on fingers, old ladies who had fallen over, dog bites, a young man who had fallen into a canal, an old man who drank fifteen pints of Guinness a day, fitted and forgot where he was. Each patient would be put on a trolley bed in a cubicle, and the white light outside would remain lit until they were seen. Six white lights created a silent pressure that was impossible to ignore. Dr R., the consultant in charge, a skilled, compassionate and paternal man, would make forays down the corridor, insisting that we took our breaks, and reminding us daily that we should all 'enjoy ourselves'.

My colleagues quickly discovered that I would see the mad, the depressed and the suicidal without complaint: a woman in her fifties whose husband was having an affair; a young woman with five children whose husband beat her; a middle-aged man whose wife said she wished he was dead, so he swallowed thirty Mogadon. Some of them were very unwell, a few of them wanted to die, all of them were using drugs that were supposed to make them feel better to harm themselves. And when the shift ended, I could leave, no bleep, no on call.

Shift work also gave me the freedom to get down to Greenham, where the women who had marched from Cardiff had now created an encampment. It hadn't been planned. Over

19

the two weeks it had taken to walk to the base the marchers had become increasingly fed up with the lack of media interest. They decided that upon arrival four women would chain themselves to the base fence and they would demand a televised debate with a government representative on the issue of cruise missile deployment. The press was indifferent to the chained women, but local peace activists in Newbury brought food and equipment and encouraged the women to set up a more permanent camp. The women decided to stay. The Ministry of Defence responded that the matter had been fully debated, citing discussions in five British universities. They told the campers that they could stay as long as they wished.

I hitched down in December to find two feet of snow on the ground and bright painted signs saying: 'Women's Peace Camp, No Cruise Here, Honk If You're For Us'. There were ten caravans and two tepees crowded into a space next to the main gate. In one of the tepees some twenty people sat around an open fire listening to a woman with a friendly face and short red hair who was talking. Helen had caught the media's eye because she'd given up five children, husband and home to live here. She was forty-four, a midwife, and 'asleep like everyone else' until she'd gone to an anti-nuclear rally in Wales. She was one of the women who had chained themselves to the fence. 'Seeing those young people trying desperately to stop their futures being damaged, I realised my generation hadn't worked hard enough to prevent what's happening. It was a fit of guilt that got me going', she told me when we chatted later that day.

Initially she'd joined the Labour Party and had stood as a councillor in Powys, her home county, to get the issue raised. Now she felt it was more important to be here:

> People are disillusioned with politics and rightly so. They need
> something they can focus their attention on and begin to believe in.
> You can actually draw a line here and make a positive physical

statement that cannot be misunderstood: we are not allowing these weapons in. If I'd said it in Powys, they wouldn't have listened.

She hadn't planned to take direct action. But, like many of today's climate activists, action came from a combination of guilt and responsibility:

> I thought the way a lot of women with children think, that I'd a greater responsibility to them than what I felt was right for me. The thought that these missiles could kill all the people I cared about anyway changed that. The only logical thing to do was to make a committed stand.

She had made the decision without consulting husband or children: 'Obviously they were very upset. The youngest is four and my husband has had to give up his job and go on social security, which he hates, though he understands what I'm doing and he supports the peace movement.' Someone handed out mugs of tea.

'How do you organise?' I asked. 'Who makes decisions? How does it run?'

'Chaotically,' Helen laughed. 'I think that's our greatest strength in a daft way, we never seem to know what we're doing from one day to another which makes it hard for the authorities to know either. People come and go all the time. No one makes rigid plans, which is excellent or you'd be stuck if something happened fast.'

'Don't you have rotas or anything?'

'Rotas?' Helen spat out the word in disgust. 'Oh my God! There were rotas but nobody paid attention to them.'

'It doesn't run smoothly all the time,' a small, dark-haired woman called Shushu joined in, 'But mostly it's OK. People simply agree on what needs to be done and do it. We meet most evenings over supper. We don't have a steering committee or hierarchy or all that kind of thing. My CND group did and spent most of its time organising jumble sales and never doing anything.'

'Why did you come?' I asked Shushu.

'I've just finished school. I've got the time and energy to do something more than organise jumble sales. Once you've accepted the fact that you can live with a greater degree of untidiness and filth you can relax. I think if we were the type of women who gave a high priority to having clean, smooth-running homes, that's where we'd be. Our priority is stopping cruise.'

A woman called Sarah walked up carrying a large puppet; an enormous woman's head with long red hair and brightly coloured hand-painted robes.

'This is the Goddess', she said.

'Right,' said Helen, 'let's walk to Newbury.' We set off, a small procession, the Goddess in the lead, bright against snow-laden branches and clear sky.[11]

The camp persisted through a bitter winter. Money, letters of support and food poured in. Newbury people stopped by the fire on their way to discos hosted by servicemen on the base; a local policeman left his phone number for anyone wanting baths. Media from Japan, Australia, the United States, Europe and the USSR all dropped by. The women who stayed impressed me. There was Franny, who ran a one-woman picket outside the construction workers' gate. Sitting with a Primus stove she'd hold out a sign saying, 'Will you stop and talk?' and another for when they returned, saying, 'How about a cup of tea on the way back then?'

Sarah had given up work in a psychiatric hostel to live at Greenham to draw attention to the silos being built to house missiles that could 'trigger off a holocaust. Every day hundreds of people drive past us and have to ask themselves "What makes those women live here and give up their normal lives?"' She saw the camp as a focus for other peace activists, uncertain what to do: 'They can come here, meet others [who are] sympathetic, get inspiration, anything could happen.'

ACCIDENTS AND EMERGENCIES

Clearly too much was happening. *Angels*, a popular TV soap opera about nurses, had characters joining the Nurses Campaign against Nuclear Weapons. Another peace camp was set up at RAF Molesworth in Cambridgeshire, the second intended cruise missile site. At the end of January 1982, Newbury District Council gave the women fourteen days to leave or they would be evicted.

The women stayed put. I caught the bus back to the camp in February to help prepare for a nonviolent blockade planned for March. I arrived to find a new marquee had been built using a poplar tree as a centre post. Banners, appliqué, poems and paintings decorated the walls and smoke rose from the central fire where Franny dispensed tea to all comers. Scores of people kept arriving. Some had come to find out what was happening about the eviction, others to train for the blockade. A friend and I ran our nonviolence training workshops in a borrowed church hall as planned. But it was clear that a major row was brewing. A week previously, all the women present at the camp had met and unanimously decided that while under threat of eviction only women should actually live at the camp. Although up to this point men had been allowed to live there, the march and camp had always been women's initiatives, with women talking to the press, representing the camp and doing the decision-making. Many now felt they should make this position clearer. Others returning the following day were unhappy with the decision, as were the men involved. By midday on Sunday, people stood shouting at one another around the main fire:

'Men are just as capable of being nonviolent as women.'

'Yes, but they often provoke more violence because the police pick on them deliberately.'

'Have one man here and it's just assumed the good ideas are his.'

'No one can ask me to leave! We weren't even allowed in the meeting when you made the decision. Anyway, this is my home—I've squatter's rights!'

'Cruise missiles are going to affect everyone, not just women.'

'Yeah, you're splitting the peace movement!'

'Do you want a lesbian separatist peace movement? Is that what you want?'

'Well! My CND group has always supported Greenham. They'd be very upset to hear you'd gone separatist!'

Voices were rising. I could not bear it, so I stood up and shouted: 'If there's ever going to be any resolution, this discussion needs a facilitator and if no one else wants the job I'll do it. I don't live here and I don't mind what you decide, but please sit down and take it in turns to speak.'

To my complete astonishment it worked. People sat down, put their hands up, spoke one at a time and actually listened to one another. There seemed to be three groups: the men who understandably resented being asked to leave; women who wanted to work in mixed groups; and the women who wanted to make it women only. This last group argued that when men were present, their views and methods tended to dominate, and left no space for women to voice different kinds of feelings and find their own ways of working. If that space was created, far from splitting the peace movement it could strengthen it by inspiring other women to take action. The women suggested that if men really wanted to support them, they could do so by giving women space. They could always start another mixed camp elsewhere. After some twenty minutes, when everyone who wanted to do so had had a chance to express their views, I suggested the women who actually lived at the camp went off into a caravan and made a decision. 'And perhaps you could do it in twenty minutes,' I added, 'so that we can hear what it is before we all have to get home.'

Thirty minutes later, again to my complete astonishment, the women reappeared. They had reached a consensus: 'Women only for two weeks.'

ACCIDENTS AND EMERGENCIES

Franny wrote to me a few days later from the peace camp, saying that 'despite the heavy scene on Sunday we now all feel whole and confident in our actions. We thank you very much for your energy and the way you brought about our decision making.' Thank you Howard, I thought. All that facilitation training had been useful.

Of course, it was not that easy. There were other women not present that Sunday, who remained deeply unhappy, especially when the camp women decided, two weeks later, that the 'Women only' rule should be made permanent. Franny changed her mind:

'I don't think we analysed the problem properly. It was those particular men and none of us had the guts to say "shape up or piss off". And when the decision was reviewed those who disagreed weren't there. It wasn't humane either. People were turning up at night in the dark and rain and having to leave. I felt we were strong enough as a group to cope with that and not stick to rigid rules.'

'So why did you stay on?' I asked.

'The women felt so strongly about it. Besides there was the blockade.'

The idea was to have the blockade following a festival that women were holding to celebrate the spring equinox. But we'd done little to prepare. I arrived at the camp sick with anxiety. 'Don't worry Lynne,' Shushu smiled, patting my shoulder, 'We're very organised.'

They were. One group of women had written a small practical briefing, outlining how the blockade would proceed, what to do if arrested, solicitors' phone numbers and so on. A system had been devised whereby every woman involved in the action would register and join an affinity group to do nonviolence training. Each gate into the base would have a legal observer who would watch what happened, note the arrests and relay information to a central point via walkie-talkies hired for that purpose. All Saturday afternoon we sat huddled in caravans getting to know one another, deciding how we'd deal with confrontation. Outside, the festival surged around us, dancers in radiation suits, jugglers, clowns, the Fallout Marching Band, all undaunted by the miserable weather. Inside, 200 women in the space of an hour managed to work out how to cover seven gates for twenty hours, sleep, eat and communicate. I couldn't risk arrest with work on Tuesday, so I was a legal observer.

I was still scared as my group walked out to fill the main gate at 6.30 p.m.: a line of police behind, an enormous crowd in front and camera lights glaring in our eyes. We sat down. At the six other gates women were doing the same thing. We waited. The rain poured down steadily but the stream of traffic was absent. The base, it seemed, in consideration of our wishes, had closed itself down. So we settled in, wrapped in rugs and macs, and took turns throughout the night to do four-hour shifts. The cameramen went home. The rain turned to drizzle. Supporters brought hot tea and entertainers left over from the festival went from gate to gate with fiddles and guitars. In the morning we discovered the police had created a new gate by cutting a hole in the fence on a deserted bit of road. 'This,' the police inspector said courteously, 'is our gate, and if you sit in it, you will be arrested.'

The women sat down. I sat in the dust at the side of the road observing. The rain had stopped. It was the first warm day of the year. Newbury Church was just visible over the horizon, the hedges and trees touched by a green mist, a lark singing, high and clear. 'Isn't this getting a little ridiculous?' the policeman asked me. 'How many of you want to get arrested?'

Ridiculous to you perhaps, I thought; to me it seems like the most sensible thing I've ever done in my life. In fact, they arrested very few. Instead, they let the traffic pile up, then swooped down and pulled women out of the way, letting them back when the traffic had passed. In response, more and more women piled into the gap. It was a wearying process. Traffic, mostly gravel trucks, came in every half hour. The women were good at going limp but it only slowed the police down. There was some rough treatment: pulled hair, a woman thrown to the ground, but on the whole the police were friendly, one even sharing his fears about nuclear war with us. At sunset the police pulled the barbed wire back across the gap; they were leaving. All the women started shouting and hugging each other: 'We've done it! We've done it.'

Singing and dancing we formed a circle. Then suddenly, spontaneously, silence fell. We stood filling the road. A policeman giggled, but then muffled his radio. They were silent too; for the first time that afternoon with us, not against us. 'Pass a smile around', a woman whispered in my ear grinning.

I whispered to my neighbour and watched as the women's faces lit up. A letter in *Peace News* the following week read:

> The blockade as a blockade was a failure and ... I would say that what I saw symbolically, and actively, manifested was woman's oppression and subjugation [...] they were using their bodies the way they've been used for centuries, lumps of unintelligent flesh booted aside when they got in the way. No attempt was made to seal off the site in a proper pacifist way with glue, padlocks and chains.

Angered, I wrote back:

> What mattered was the effect that symbol had on ourselves and the people we were confronting. A policeman trying to pull two arms apart in a firmly linked human chain has to directly confront his own feelings about human contact, handling women not as sexual objects but as powerful beings [...] hacking through a chain he can avoid all that. It's just the nice easy masculine field of mechanics, no feelings involved. It's that objectification, that lack of understanding as to what pain and suffering really mean that makes it possible to press a button and annihilate a million people, that's what we're trying to

challenge. That action meant directly confronting these people with our bodies [...] with the comprehension of what violence and power mean in human terms [...] No, we didn't 'win', but you can't take control till you feel powerful and not one of us left that day without feeling stronger and more sure of our power to act. Surely not the result of a day of oppression and subjugation.[12]

Two days later I handed in my resignation. After six months in adult A & E, I was in my second month of paediatric A & E and was just beginning to feel useful. I could doggedly sew up cuts above the eyebrows while a small form struggled in their parent's arms and went red in the face with rage. I could recognise a greenstick fracture and was no longer too alarmed by temperatures over 40°C, knowing how to check for the tell-tale signs of meningitis. I knew I was making a difference to individual children's lives, but still, it did not feel like enough.

This was the first job in which I had a female boss: an extremely competent paediatric A & E consultant with a no-nonsense, get-on-with-the-job manner that reminded me of my physical education teacher at school. She quickly glanced through the resignation letter over which I had agonised for two days, and stared across her desk at me with a mixture of pity and bewilderment.

'You do understand this will mean the end of your medical career?'

'I am sorry to hear that. But I don't think I have a choice. I'm completely exhausted from trying to do two jobs at once. I don't want to do this one badly or put patients at risk but I cannot give up the peace movement. I think the threat of nuclear war is a real emergency.'

'How will you live?'

'I have some savings', I mumbled. I had not really thought through that part.

Ms O. sighed, put my letter aside and shrugged.

'It's up to you of course.'

All week colleagues and friends continued to echo her warnings, telling me I was 'a silly girl', that I would 'lose all credibility', and that 'no one will want to risk offering you a job after this'. I sat on the narrow bed in my tiny, gloss-painted bedroom, staring at the streaky, beige linoleum floor. Perhaps they were right. Perhaps this was the end of my career. But perhaps I was entirely unsuited to the medical profession anyway. A medical career was not just about helping sick people recover but about fitting into a complete world. These years as a house officer had shown me how badly I fitted, even before I got caught up in the peace movement. I played bridge, tried to join in discussions as to which was the best social club, best car, where to go on holiday, who might marry who, and the hilarious behaviour of the patients. But I was not convincing. A mild infatuation with an obstetric senior house officer with striking grey eyes only lasted a month, because 'You are too serious, too intense. You make everyone feel they have to justify themselves.'

I had the constant feeling I was play-acting, putting on nylons and a skirt and blouse; hair neatly combed; make-up not too much, but not too little; speech not too soft, or consultants would ask you to 'speak up', but not too loud, you did not want to be thought aggressive. I had a whole list of rules for behaviour: Be careful not to challenge. If your boss in the pain clinic tells you that there are five grades of pain with aneurysms, cancer and cholecystitis at one end and period pains at the bottom, avoid asking where he would place labour pains. Bite your lip when one male colleague tells another to avoid a particular X-ray technician because she is not being helpful and probably has PMT. Just smile politely when a senior registrar, training you and your colleagues in plastering broken limbs, says: 'Now we are going to ask Dr Jones to take off all her clothes, so that we can use her as a model.'

SORRY FOR THE INCONVENIENCE

The trouble with long-term lip-biting and polite smiling is that at some point it all becomes too much. Near the end of my time in adult A & E, one of my fellow junior staff, Dr Q., gave us a lecture on the knee, and the numerous ways in which it could be injured. He decided to make it more entertaining by throwing in some *Playboy* pictures of naked young women from the rear, bent over, showing not just their knees but everything else. Everyone sniggered, while eyeing me to see how I would react. 'I find this really offensive', I murmured, for once quite unable to bite my lip.

Everyone roared with laughter: everyone being fifteen other young men, two male consultants, two male students and four other women doctors. This was just what they expected. I lost my temper:

> No, you are not going to laugh, you are going to hear me out for once. What do you think I get out of complaining? Nothing! Do you think I like being the token feminist endlessly laughed at? I don't. Why should I have my body up there on the screen displayed in such a degrading way? I find it humiliating and disgusting, and think about how it makes you view the actual woman with a painful knee when she is lying on the couch? I know it's hard for you but just for once in your lives try and put yourselves in our position!

I was too upset to continue. I burst into tears and got up to leave the room. One of the consultants got up, as I stumbled out between the chairs: 'Before you go, Lynne, I would just like to say I am in total agreement.'

I left the room in a stunned silence. As usual I was furious with myself. My tears and emotions would have devalued the argument in their eyes. Why couldn't I keep my cool? I pulled myself together and returned, sitting down at the back. People kept looking at me nervously, like a bomb that might go off any second. Dr Q. came and apologised at the end of the meeting. Dr R. summoned me for a little chat in his office the next day.

He told me that he admired my bravery, but I would never change anything, and he did hope that I was not unhappy.

'Perhaps you are too critical when really all these boys are very compassionate types. It is just a bit of fun you know.'

'I am well aware of their compassion, that's not the problem. It's their view of women as primarily sexual objects that bothers me.' I was getting teary again and Dr R. noticed.

'Well, if you are going to get emotional there is no point in discussing it.'

How was I to explain how wearing I found the 'bit of fun', when it went on every day and coloured so many exchanges; how exhausting it was to be silent, unemotional and invisible? Possibly female doctors did not really exist. After all, house officers were 'housemen'. 'Housewoman' was not a word, and you only had to glance at the adverts in a medical journal to see that doctors were men, and women were nurses or patients, usually suffering from depression or anxiety. In my first week in A & E, I had walked into a cubicle with a male medical student following me, and the patient started to give him the history, as if I was not there.

'Umm, hello sir, actually I am the doctor', I said.

'Oops, sorry miss', the man said, winking at the student.

Increasingly silence felt like a kind of collaboration. I thought back over all the times I had just sat there while male consultants shouted at nurses and secretaries and their junior female staff. One of the consultants at medical school would have a temper tantrum if the patients were not suitably undressed, ready to be examined, as he did his ward round.

Only a month previously I had sat in the orthopaedic clinic watching Mr M. examine an elderly lady with a painful knee. He was rapid, thorough and ruthless: she yelled as he pressed on the patella: 'Right, I know the diagnosis—that's the good news.' She winced as he pressed again, disregarding her and talking directly to me: 'The bad news is I cannot do anything about it, she will have to put up with it.' He rushed out, having said not a word to the patient, muttering to me: 'We all get pain in the knee now and then, it would be a major operation to remove the patella, not justified. Stupid complaint, stupid woman, not worth the trouble.' My biggest fear was of developing a similar indifference.

The Medical Campaign against Nuclear Weapons was not a refuge. At our medical conferences and meetings, I noticed the speakers were mostly men, while the audience was mostly female. But when I had commented, at one executive meeting, that we were as patriarchal and hierarchical as our profession, an eminent female professor told me:

'One struggle at a time is enough.'

'You do your cause and yourself a disservice', another male friend told me.

Yet I meant what I said in that speech I gave in Bristol. Challenging and ending patriarchal politics was fundamental to getting rid of nuclear weapons, and addressing other global problems as well. I was not doing very well trying to challenge sexism within the medical profession. Perhaps working for peace with other stroppy women camped outside a missile base was the right path for me. I knew it looked ridiculous. And if we failed, what then? But what if we succeeded?

Helen rang me up.

'Lynne! You can't give up your job! We need doctors.'

'I consider nuclear weapons the biggest health threat and I'm not confronting it here in casualty. You're a fine one to talk anyway. Have you been evicted yet?'

* * *

Diary: May 2019, St Ives, Cornwall

Not again. I really don't want to be here, sitting in a Methodist church hall, listening to a speaker trying to terrify me into taking action against an existential threat. The church is rather similar to that one in the cathedral town forty years ago. Late evening summer light filters through the Victorian Gothic windows. The dark brown varnished wooden pews are full of people of all ages. My husband Asmamaw and I share a pew with a retired professor of philosophy and a trapeze artist in her twenties with dreadlocks. Dom, the speaker, is a sustainable builder from St Just. He has none of E. P. Thompson's charisma; indeed, he tells us at the beginning that he hates giving this talk, but we have to face the facts and they are profoundly upsetting. Yet his quiet, unassuming manner holds your attention. As does the reality he shares with his PowerPoint slides, reminding us that 200 species are being made extinct every day, that on average 60 per cent of all vertebrates have been wiped out since 1970, that bees are dying and coral is bleaching, and that there is likely to be no ice in the Arctic in five years' time. As for Ethiopia, Asmamaw's homeland, Dom puts up a map: a projection of a world at 4°C warmer. It is predominantly yellow and brown. Three quarters of the world's population currently live in the central latitudes that will experience deadly heat. These regions, including most of Africa, all of India, China, Australia and the United States, will be uninhabitable. Any survivors will have to flee to join, or more likely fight, with what remains of humanity, crowded into the green zones of western Antarctica, Canada, Russia, Scandinavia and Britain.

As he talks, I find my mind wandering back. Here is the strange thing about these times. The threat of nuclear weapons has not gone away. As treaties unravel, the possibilities of nuclear conflict through accident or design grow. But it is still a future threat, nuclear weapons have not been used since the end of the Second World War. Climate disruption and the destruction of the natural world is happening right now. Last week the concentration of carbon dioxide in the atmosphere exceeded 415 parts per million. The last time levels were this high, there were beech trees growing in Antarctica, Homo Sapiens did not exist and the site of this small town was under water.

So, what to do? Forty years ago, when my boss told me political protest would mean the end of my medical career, I was still uncertain if I wanted one. Somehow, more through luck than judgment, I found my place, as a psychiatrist establishing mental health programmes in humanitarian emergencies. Yet it is the climate crisis that has created many of the front lines where I have worked, contributing to, and in some cases causing, the conflicts and disasters that have driven people from their homes: from droughts in East and West Africa, hurricanes in Louisiana and the Philippines, and now with migrant crises across Europe.

Sometimes I wonder if my direct engagement in humanitarian work has provided an excuse for less action in other areas? There is no question that providing immediate relief to an individual is simpler than trying to unravel and address the underlying problems that caused the suffering in the first place. Every charity and NGO will tell you it is so much easier to mobilise resources and people for response rather than prevention.

And yet, as disaster follows disaster, I am no longer certain how good we are even at simply responding. Two cyclones hit Southern Africa this spring devastating the lives of more than 2 million people, mostly in Mozambique. This followed last year's drought that had already left the same number without adequate food. The new crops about to be harvested were destroyed; villages including health centres and schools were flattened; water supplies cut. Cholera threatens, and some 70,000 pregnant women have no access to safe deliveries.[13] *I stare at the photographs of Ibo, the island where we lived and worked through 2013, and see nothing but destruction: mud, ripped palm trees and the outlines where houses once stood. Ninety-five per cent of the buildings have been shattered to their foundations, and on that small island alone, some 6,000 people still sleep in makeshift shelters in the rain that continues to fall. It is the same situation along forty miles of coast.*[14]

There had never been a cyclone in this part of the country before. Kenneth increased from a Category 1 storm to Category 4 in one day and dumped an unprecedented amount of rain. Climate scientists all agree that climate change allows storms to hold more moisture and contributed to this devastation. These events caught the media's attention for some forty-eight hours. We are too preoccupied with Brexit. Yet Britain's vote to pull up the drawbridge was in part a response to the migration crisis. And if the response across Europe to 1 million refugees is to retreat into tribalism, how will we respond to the possible 1 billion that the UN predicts by 2050?[15]

In 2015 The Lancet *set up a global collaboration to monitor and report on health and climate change. In its report of November 2018, it stated:*

> *Trends in climate change impacts, exposures, and vulnerabilities show an unacceptably high level of risk for the current and future health of populations across the world. [...] Given that climate change is the biggest global health threat of the 21st century, responding to this threat, and ensuring this response delivers the health benefits available, is the responsibility of the health profession; indeed, such a transformation will not be possible without it.*[16]

A video circulates on Twitter. Dr Alex Armitage, a children's doctor working in an Accident and Emergency department, believes the world is facing 'drought, mass starvation and societal collapse'. He argues that 'as a paediatrician I am one of the people who is responsible for the health and wellbeing of future generations', and that 'the medicine that the government needs right now is [...] mass participation in nonviolent civil disobedience'.

Dom has come to the end of his talk.

'To change everything, it takes everyone', he remarks.

'That's us', Asmamaw says.

LETTER TO A WOMAN ON THE FALKLANDS VICTORY PARADE, OCTOBER 1982[17]

I don't know your name, in spite of having stood beside you all morning on the parade. I never discovered it. But your face stays clearly in my mind—uncomprehending, distressed. It won't go away. That's why I wanted to write—to try and explain. We'd known about the parade for weeks; known that while ostensibly it was saying 'welcome home' to the soldiers, it would also be saying 'war is glorious' and while saying 'well done boys', it was also praising Mrs Thatcher for sending them to war. We knew that while the hardware, the guns, the tanks, the marching bands would be there for all to see, there would be no coffins, no victims of burns, no drowned sailors. So, we had to go ourselves—to say there's nothing glorious about war, and nothing well done about problems solved by killing people. (That's if you can call the problem of the Falklands solved in any way; I still feel sorry for those people living in what has basically become a military garrison.)

We're just a small group of women. We knew it wouldn't be easy to make our point clearly and without offence. For you have to believe from the start we didn't want to offend anyone. It is war and those who cause it that offend us. So, we thought we would do it very simply. Go as a group, and in silence turn our backs on the parade, holding up a banner saying 'Women Turn Their Backs on War'. That was all. We knew that people around us would be hostile, that they would shout and perhaps physically abuse us. Such occasions seem to breed a violence of their own—where no point of view but that of the mob, even one peacefully expressed, can be tolerated. Odd, isn't it? Because the Falklands victory was surely, if nothing else, victory for our 'democratic liberties' including that of freedom of speech—something sadly missing in Argentina.

Well, we prepared ourselves. Some of us took the parts of hostile onlookers so that the others could experience what it would feel like. We organised ourselves so that some would form a cordon of peacekeepers round the actual demonstrators, giving them at least some protection. I was one of these—my job to talk and pacify if possible. We hadn't prepared ourselves for you however. A plump, smiling woman, your hair freshly done, bright blue eyes, who came and stood right in the middle of our group. You chose us deliberately, you told me, because we weren't too tall and you thought you could get a good view over our shoulders. It had been a difficult morning before you arrived. We had got there early to get a good place by the barrier. The women holding the banner however had got the sections muddled and had to go up and down to the underground station to get it sorted out. The coming and going alerted the police who got suspicious and searched us. They didn't find the banner (we wore it under our clothes) but they found a leaflet in my bag with the giveaway words 'Feminism and Disarmament' (are they really so frightening?). Anyway, they warned us they'd keep their eye on us. Any barracking the troops or jumping the barriers and they'd be down on us like a ton of bricks. Of course, we could promise them there would be neither.

It began to get crowded. And you arrived. Friendly from the start, you told me you were worried your camera wouldn't work—you didn't know how to use it, did I? Or perhaps one of us could take the pictures from the front. Your son would be marching by and you didn't want to miss him. That was when it hit me. How really difficult, almost impossible it was, what we were trying to do. Here you were, in your best clothes, come a long way with your husband to see your son, who'd got home safe from the war, have his moment of glory. Little enough reward for having put up with the horrors of the South Atlantic. And here was I, equally glad your son was safe, and wanting to deprive him of that moment—seeing in it the seed of other wars, from which he might not come back. How could I explain that to you, standing there so pleased and so proud? How could I explain that I thought we had to stop being pleased and proud and be bitter and sad and angry and say, 'It has to stop. There must be another way, some method of solving conflicts that doesn't waste and destroy. Finding it would really give us reason for pride.'

I had two hours before the parade came by. If I had tried, would you have listened? Or would you have called the police and had us removed? I don't know because I'm afraid I lacked the courage. I said nothing, but I couldn't bear the thought of how you would feel when we turned around and held up our banner. I didn't want you to miss your son. I asked the woman at the front to find a place for you. You wouldn't take it at first, insisting we were 'too kind', but you gave in and moved next to Karen. I felt like Judas. The waiting was awful. I saw you chatting and making friends with Karen, everyone does. Your place beside me had been taken by a severe-looking woman in a smart blue hat who eyed me suspiciously and made no attempt to talk. Mary, next to me, was having problems with her child. The man behind her didn't like her baby carrier. But his wife told him to shut up and told us not to mind—she had grandchildren herself. The child went to sleep. It started to rain. A band somewhere was playing 'hits' from 'The Sound of Music' and 'My Fair Lady'. We sang along in a desultory sort of way.

Then the helicopters came. Did you see them? What did you feel? I can only tell you that to me there is nothing beautiful about helicopters. They lack even the rudimentary bird-like grace of an aeroplane. And as they loomed over us, black insects with white search-lights for eyes, startling the birds, I felt only a cold gripping sensation of fear—whilst around me the crowd cheered and cheered. The lady on my left was waving her Falklands souvenir programme and the expression on her face was rapt. And then the parade itself began. We let one contingent go by and then seven women in front of me turned—arms holding up the banner, quiet, quick and simple. And the crowd went wild. The hard-faced woman beside me lunged for Andrea's piece, screaming, 'I suspected you from the first', pushing and thrusting. We let her through to the barrier. The police grabbed other pieces and stood solid the other side. The man behind us was yelling 'you bastards' and another was screaming about 'our country' and amid all the hubbub I saw your face turn to Karen and I saw tears pouring from your eyes. 'I thought you were my friends! I thought you cared about my son. Don't you dare upset me like this. Don't you dare upset me!' Karen was crying too, her back turned to the parade and the policeman shouting over her head, 'Ignore them, ignore them'. And I, forgetting my peace-keeping role, shouting to make my voice heard, said to you, 'We're just as upset as you. It's because we care about your son that we're here!' I wanted to say more, but you turned away to the woman on your other side for comfort, and then a policeman lunged over the barrier at me, hands around my neck, saying, 'Right, you dirty cow. I'm taking you to the police station where scum like you belong.'

FALKLANDS VICTORY PARADE, 1982

So, I didn't see the rest of the parade, though I know the rest of the women stayed, backs turned and silent to the end. And of course, I didn't see you again. Which was a pity, because we share the same values, you and I. We love freedom, and happiness. Only perhaps you would tell me such things can only be maintained because your son fights to protect them. And I would reply: the fact that he has to fight destroys those things in themselves.

2

I WISH THERE WAS SOMEWHERE
WE COULD MEET

It's the woman in the red anorak who sticks in my mind. She is young, has long dark hair and is completely furious. She spits on the ground in anger and disgust and walks away. It was October 2019. Asmamaw and I were now part of a local affinity group in Penzance called Dolphins, because we met weekly in The Dolphin pub. We had decided to answer Extinction Rebellion's call for mass action in London and had arrived in Trafalgar Square to find it transformed. Instead of traffic grinding past Charing Cross station and down Whitehall, large groups of people were sitting at the road junctions, chanting and singing, gathered around protesters who were glued to the road or each other, or attached to strange cubic tower-like structures.

Affinity groups from the South West met by one of the fountains and quickly reached a consensus to go and protest outside an oil and gas conference taking place in a hotel in Liverpool Street. Before leaving, Asmamaw and I checked in at the Wellbeing tent set up in the Square to collect coloured sashes that identified us as outreach and wellbeing. For the moment, we had both decided we were more use supporting those who were prepared to be arrested if necessary, making sure they were physically and psychologically OK, rather than getting arrested ourselves. I already knew to my cost how a criminal record could affect our ability to obtain visas to travel and practise abroad, and neither of us wanted to give up doing relief work when it seemed certain more would be needed.

A large crowd of protesters was already seated on the street in front of the hotel entrance on Liverpool Street, and a line of police blocked the door. No one was being arrested and the police had helpfully parked a car across the junction with Bishopsgate. Then half the crowd started 'swarming'. I had not seen this before. It involved a crowd walking out onto a traffic junction at the red lights, filling it, holding up their banners and remaining after the lights changed, but only for five to seven minutes. Then they left to allow the traffic through, and started out again five minutes later. The idea appeared to be disruption, but not too much. The traffic built up in seconds and a blonde-haired man in the large white van at the front, leant out, waving his fist at us:

'I have to go to a fucking job, I've got hot food in the back, what are you fucking doing?'

'I can explain', said Deepa, a doctor friend who was handing out leaflets with 'I'm sorry' in large letters on the front.

'I don't want to hear! Get a fucking job.'

Deepa moved on to the next car. Astonishingly the policeman standing beside me, who was doing nothing to clear the road, joined in the explanation.

'It will only be five minutes, sir.'

'I am really sorry for your inconvenience, we are trying to prevent a much bigger disruption in the future', I added.

'I don't care! Get a fucking job!! Why aren't you at work?'

'Because I want your children to have a future.'

'They have a fucking future; they are at work.'

While I was standing there another man, large, balding, in a red tracksuit top walked up and accosted me directly: 'You are all the same! You're against Brexit, against Trump, you should listen to Nigel Lawson. He explained it all. The Thames froze over 1,000 years ago and we are all fine!'

A black cab was trying to do a U-turn just as the blockade cleared and traffic started moving. When the blockade reformed, I grabbed some more leaflets and worked my way up the new line of traffic, mostly delivery vans and buses. A surprising number of people were positive, smiling, nodding and giving a thumbs up and taking the leaflet. 'You got to inconvenience people sometimes', said a man in a woolly hat with an earring. A minicab driver was less happy. 'Look love, I got Greenpeace stickers all over my car but please, I gotta get through!'

By mid-afternoon there were around 500 activists in the area and the swarming had transformed into a permanent blockade across Liverpool Street and Bishopsgate. Many were simply sitting in the road; some were glued to a tin drum. A large circle of drummers danced around them. The traffic had been diverted and the roads were now completely clear except for parked empty buses. All those prepared to be arrested seemed comfortable and content, and the police were still not making arrests, so I decided to leaflet passers-by.

I had now acquired a neatly made placard with the brightly coloured Shell logo, with the 'S' deleted, so it said 'hell'. I stood on the broad pavement with the placard in one hand offering leaflets with the other. I noted six types of responses:

1) Walk past with fixed blank stare in other direction (both genders but more men than women);
2) Glance at me, look anxious, look away (mostly women);
3) Glance at me, look hostile, walk on (mostly men);
4) Talk earnestly to mobile phone (both genders—younger);

5) See me, smile, take leaflet (no one type, but I noticed that most people of colour fell into this group);
6) See me, look slightly curious.

The last were the group with whom I tried to engage. There was no point upsetting the anxious or the hostile, and the smilers probably already supported us.

I had also tried various opening sentences to catch their attention: 'Good afternoon, I am a doctor—this is an emergency' did get attention, but then a head-shake and reversion to one of the above. 'Good afternoon, Madam/Sir, can I explain what we are doing?' was no good—they had already passed by.

Saying 'You look like a very intelligent person', with a big smile, worked astonishingly well, at least for a surprising number of men—even some of the very large, tall ones with the fixed, blank, 'I am not seeing you, you don't exist' stare. Yes, flattery will get you everywhere.

Three immaculately suited American men coming out of the conference saw my placard and asked to take my picture: 'Only if you take a leaflet', I replied.

It was then that the woman in the red jacket approached me, or rather us, because at this point another young woman leafleteer had joined me. The woman walked up to us directly: 'You are stopping me getting to my child on time. She's only eighteen months and I am going to be late!'

The woman was screaming and furious, real hatred in her eyes. My young friend offered her an 'I'm sorry' leaflet.

'Don't give me that! Do you understand? She's eighteen months! What do you think you are doing harming her?'

'I am so sorry, it's because we don't want her harmed in the future', I began.

'Don't give me that! Do you fly?'

'As little as possible, and only for work.'

'Do you drive a car?'

'Again, as little as possible, I live in a rural area.'

'You're fucking hypocrites all of you, blocking my way, making my daughter unhappy, stopping her seeing her dad.'

She walked past us muttering angrily, and as my companion shouted after her 'I am so sorry' one more time, she turned back and stood still, then spat dramatically on the ground in our direction, and walked away. And I was back on the Falklands Day parade in October 1982, not so far from here. Once again, I felt completely overwhelmed by the impossibility of what we were trying to do, and of finding the right way to do it. Dialogue or disruption? Is it possible to have both?

* * *

In the spring of 1963, Martin Luther King Jr and the Southern Christian Leadership Conference, at the invitation of Fred Shuttlesworth and the Alabama Christian Movement for Human Rights, embarked on a campaign for racial integration in Birmingham, Alabama. King described the city as 'probably the most thoroughly segregated city in the United States'.[1] The campaign involved the key components of what is now commonly termed civil resistance. These were, first, a commitment to nonviolence: participants would not inflict violence on others; second, sacrifice: the participants were willing to accept hardship themselves in the form of violence dealt out to them, including arrest and jail sentences; and third, disruption: the actions would disrupt business as usual and thus create pressure for change. Over a number of weeks, Project C, as the campaign was called, combined boycotts of white businesses that segregated customers, with sit-ins at segregated lunch counters, kneel-ins at white-only churches and with marches by children and young people in defiance of an injunction on such protests. The pictures of notorious Bull Connor, Birmingham Commissioner of Public Safety, using high-pressure water hoses and police dogs on young non-violent protesters drew international condemnation and nation-wide support for the protesters. Meanwhile, the local jails were filled beyond capacity with protesters, and the economic effects of the boycott, combined with a downtown area under siege, were being felt by local businesses, who started to put pressure on local government.

On 10 May, after a moratorium on demonstrations during a period of negotiation, Birmingham City Council agreed to desegregate lunch counters, restrooms, drinking fountains and department store fitting rooms; to hire Black staff in stores; and to release the jailed protesters on bail. It was a notable victory, although tragically, segregationists responded to this agreement with violence, including the notorious bombing of a church in which four young Black girls were killed.[2]

The campaign also resulted in one of the most famous and seminal documents for nonviolent activists: the *Letter from Birmingham Jail*, which Martin Luther King wrote while in jail himself. It was a response to eight white religious leaders, both Jewish and Christian, who had written an open letter to the *Birmingham News* condemning the nonviolent actions, which they called 'extreme measures [...] led in part by outsiders', as both 'unwise and untimely'. In an earlier letter they had made clear their opposition to civil disobedience, arguing that disagreement is not a reason for 'advocating defiance, anarchy and subversion' and that change should be achieved through negotiation and the courts.[3]

King's response has provided some of the most powerful arguments for nonviolent direct action as a method of achieving change. He pointed out that he was there by invitation, but that in any case, 'I cannot sit idly by in Atlanta and not be concerned about what happens in Birmingham. Injustice anywhere is a threat to justice everywhere.'

He explained that having collected the facts, negotiation had been tried and failed, and so they had no choice but to 'present our very bodies as a means of laying our case before the conscience of the local and national community'. He argued that:

> Nonviolent direct action seeks to create such a crisis and establish such creative tension that a community that has constantly refused to negotiate is forced to confront the issue. It seeks so to dramatize the issue that it can no longer be ignored. [...] We must see the need of having nonviolent gadflies to create the kind of tension in society that will help men to rise from the dark depths of prejudice and racism to the majestic heights of understanding and brotherhood. So, the purpose of the direct action is to create a situation so crisis-packed that it will inevitably open the door to negotiation.[4]

Roger Hallam, an organic farmer from West Wales, and one of the founders of Extinction Rebellion, gave those actions in Birmingham in 1963 as examples when he came to talk to a packed school hall in Penzance is the summer of 2019. He had begun with a darker and gloomier version of the talk Dom had given that May, warning us that the likely loss of permanent Arctic sea ice and subsequent disruption of the Gulf Stream in the near future were irreversible tipping points that would cause chaotic weather over the coming years; while the degree of global warming[5] already locked in meant that social collapse, food shortages, mass starvation and war were almost inevitable: 'I don't grow food outside anymore. There are hundreds of farmers shitting themselves every year because it's a casino.'

So, we really had no choice but to act. But going on marches, signing petitions and lobbying MPs had all been a catastrophic failure. Half the carbon emissions we were dealing with had occurred since Al Gore's showing of *An Inconvenient Truth.*[6] As Hallam concluded, 'What really works is causing a fuss.'

By which Hallam meant two things: disruption and sacrifice, because 'Without suffering people will not take you seriously, there is no change.' He used as an example his own hunger strike to get King's College London, where he had studied for a PhD on radical social movements, to agree to divest from fossil fuels in 2017. Now he was arguing that large numbers blocking the streets of a capital city, filling the jails and thus overwhelming police resources, would force the government to respond.

He based this view in part on research conducted by the political scientist Erica Chenoweth. In collaboration with Maria Stephan, Chenoweth had studied 323 mass movements aiming to achieve country-level changes between 1900 and 2006, and showed that those that relied primarily on nonviolent resistance were more successful than those using violence. Chenoweth also found that 'no revolutions have failed once 3.5% of the population has actively participated in an observable peak event like a battle, a mass demonstration, or some other form of mass non-cooperation'.[7] This figure of 3.5 per cent is much quoted and is still central to Extinction Rebellion's strategy. As they say on their website under the heading 'Act Now': 'Our strategy is therefore one of non-violent, disruptive civil disobedience—a rebellion. Historical evidence shows that we need the involvement of 3.5% of the population to succeed—in the UK that's about 2 million people.'[8]

However, Chenoweth points out that the 3.5 per cent is not a prescriptive statistic: 'Trying to achieve the threshold without building a broader public constituency does not guarantee success in the future.'[9] Active participation by 3.5 per cent is unlikely to occur in a society unless the majority of the population are already in some sense sympathetic to and supportive of the activist cause: 'Organizing only to achieve mass participation benchmarks may create a loud but wildly unpopular minority, with little chance of achieving a sustainable victory.'[10]

In a recent essay in Greta Thunberg's *The Climate Book* (2022), Chenoweth is even clearer: 'Removing a single hated dictator is much easier than agreeing to and installing wholly new political institutions, social practices and economic markets simultaneously.' She shares additional research showing that the 'critical tipping point for changing everyone's behaviour is a committed minority of 25% [...] If 25% of the population visibly change their practices, norms and behaviours, the climate movement's victories should be more widely accepted, durable and effective.'[11]

Arguably, to begin to effect that level of change the manner in which an action dramatises a problem and communicates with the public is just as important as large numbers of arrests and the disruption caused. Does the action highlight the problem and create a dialogue within people's minds about why something is happening? Black people kneeling in white-only churches and sitting at segregated lunch counters in downtown restaurants in Birmingham, Alabama, directly targeted the unjust and discriminatory laws, holding them up for examination. The actions both symbolised and dramatised what the activists hoped to achieve: justice and integration. The violence meted out towards them for doing this made visible the unjust violence of segregation occurring every day. The Birmingham campaign put civil rights on the national stage.

My own experience at Greenham also taught me that the communicative aspects of an action can be as powerful and significant as the disruption, particularly in helping a movement to grow in its early stages. This was exemplified by the 'Embrace the Base' action that took place in December 1982.

The idea for that action developed slowly over a series of meetings at the camp and in London. We knew we wanted to call a big women's action. Some of us, including Helen and myself, pushed for this to take place on 12 December, the anniversary of the decision to deploy the missiles. There were arguments about what kind of action: everyone wanted to surround the base, but that would need at least 16,000 women: how would we reach them? At one London meeting we came up with the slogan and drafted a letter: 'Embrace the Base on Sunday—Close the Base on Monday'. After more meetings and more changes, a final draft of the letter went out in mid-October. We asked every woman who received it to copy it and pass it on to at least ten others:

> The Peace Camp has been a Women's initiative. Reversing traditional roles, women have been leaving home for peace, rather than men leaving home for war ... We cannot stand by while others are organising to destroy life on our earth ... we have one year left in which to reverse the government's decision about cruise missiles. There is still time to stop them. We are inviting women from all over Britain, Europe and the World to come to Greenham Common on December 12 and 13 to take part in a mass action that will express the spirit of peace and the politics of peace ...
>
> EMBRACE THE BASE ON SUNDAY ... CLOSE THE BASE ON MONDAY
>
> Women are asked to bring personal things that represent the threat of nuclear war to us and that express our lives, our anger and our joy ... we want to decorate the entire fence with personal things.
>
> THIS IS A CHAIN LETTER WITH A DIFFERENCE. WE'LL MEET AS A LIVING CHAIN ... PLEASE TELL EVERY WOMAN THAT YOU KNOW.[12]

Thus the chain letter itself began the process of symbolisation. And there was something about it: not a poster, not a leaflet, just a request circulating woman to woman. I knew the letter was working when my mother's phone started ringing off the hook. I had offered to help with nonviolence training, and having no home of my own except the camp, had rashly put her number on at least one version of the letter. From October of that year women rang all the time, about everything.

Disruption also helped. Two small actions, one in which women had occupied a sentry box just inside the base and another in which they had tried to stop construction work, had resulted in twenty-three women going to prison within a three-day period that November. The press began to take an interest in these Peace Women. The result was astonishing. On 12 December 1982, women came from all over. The missile base fence at Greenham became the most extraordinary visual statement about what mattered in life. Women hung photographs of children, partners and grandparents; wedding dresses, baby clothes, an entire china tea set. They wove peace symbols into the wire and painted artwork. By mid-afternoon, the base was entirely surrounded by women, while TV news that night showed it encircled by candle light. *The Daily Mirror* gave us the entire front page the following day, with the headline 'PEACE! The Plea by 30,000 Women who Joined Hands in the World's Most Powerful Protest Against Nuclear War', while *The New York Times* headlined: 'Women in England Rally Against Missiles'.[13]

Meanwhile, some 2,000 women who had never met before prepared for direct nonviolent confrontation with the nuclear state by getting to know one another and forming small affinity groups. We spent the night in cold, wet tents discussing why we were there, what we were afraid of, how we would deal with any violence. And the next day women just flowed around the base like water, sitting wherever they were needed, being dragged away by the police and rolling, crawling or walking back. More women kept arriving and joined in, a massive sea of women hugging, holding and supporting one another. The police had a 'no arrest' policy, which did not stop them roughly manhandling us, but I did not see a single woman freak out or respond with violence. That night Helen and I were asked to come and discuss cruise missiles with Labour Party deputy leader Denis Healey on *Newsnight*.

A couple of weeks later, on New Year's Day 1983, more than forty women climbed over the fences and got on top of the half-prepared silos, where they danced and sang for more than an hour before being arrested and removed.

The actions themselves dramatised both what we wanted and what we opposed: a carpet of women huddled together on a darkened road in front of a bus; a circle of women in the light of dawn dancing on a missile silo. These images of life against death captured the public imagination and arguably did as much to alert the public to the dangers of cruise missiles as the actual disruptive effects of the blockade. In spring 1981, 56 per cent of women and 43 per cent of men opposed cruise missiles. A Marplan poll taken after these actions in early January 1983, showed that 67 per cent of women and 55 per cent of men now opposed the government's decision to allow the Americans to put cruise missiles on British soil. Many cited the Greenham Common women as a factor affecting their opinion. Some of those women were Tories. One of them explained: 'Before Greenham Common I didn't realise that the Americans had got their missiles here [...] it was the fuss the Greenham women made that made me realise.'[14]

That is not to say that disruption was not important, particularly later, after the missiles had arrived (as discussed in Chapter 6 of this volume), but that disruption was itself facilitated by the network of support that started to build after these actions.

Symbolism matters. As James and Ruby argue in their essay 'Cultural roadblocks' in *This is Not a Drill: An Extinction Rebellion Handbook*:

> An object placed in the middle of the road is an obstacle [...] a large concave frame of reclaimed wood in the middle of a bridge is just that [...] until it's activated by skateboarders, at which point it becomes a mini ramp, a meeting point, a media focus, an authentic creative vision of how much better things can be—a space for cars transformed into a people first festival.[15]

And that's the point: such disruption at its best might possibly 'demonstrate the future we want to see', as other essayists in the book argued.[16] These authors state that this kind of visioning, along with disruption and outreach, are the three main aims of protest, although in any particular action just one of these may take precedence.

This is Not a Drill was published shortly after April 2019, when Extinction Rebellion's visually powerful and disruptive occupations of major sites in Central London had resulted in 1,000 arrests and negotiations with the government. But it's possible that those who came up with the idea to block the Docklands Light Railway at Canning Town during the October 2019 rebellion had not read the XR handbook. I was back in Cornwall on 16 October when I woke to find a major debate going on across all my chat sites because a small group of Extinction Rebellion activists proposed disrupting the Docklands Light Railway (DLR) in the Docklands area of London the following day. My own affinity group and large numbers of people in Extinction Rebellion were opposed for a wide variety of reasons, all of which made sense to me: the action would target and inconvenience people we wanted to support; it might terrify some people who had previously witnessed horrible events on the Tube; and surely tube trains and the light railway were a method of transport we wanted to promote. My own leafleting experience over the previous days had shown me you need to offer people the option of whether to engage or not. In a crowded tube station, people waiting to get to work wouldn't have that choice, and the anger and claustrophobia created in and around stopped trains might well provoke violence.

There was something extraordinary about watching digital democracy in action in real time, and participating through online chats on WhatsApp and Telegram. By the end of the day, online polling of Extinction Rebellion supporters showed that more than 75 per cent opposed the tube action. Apparently, this was fed back to the relevant affinity groups.

Meanwhile, a couple of days earlier, the Metropolitan Police had cleared all the tents from Trafalgar Square and taken the unprecedented step of imposing a Section 14 order on the whole of London, making it a no-go area for public protest. In response, crowds of protesters, MEPs and celebrities had turned up to reclaim Trafalgar Square. Writer and environmental activist George Monbiot got himself arrested in a blaze of publicity alongside Jonathan Bartley, the co-leader of the Green Party; and the mayor of London, Sadiq Khan, criticised the Met for shutting down the right to peaceful, legal protest. A familiar dynamic was at work: if meaningful nonviolent protests are prevented and shut down, this tends to generate more protest.

Sadiq Khan also condemned the tube action; and the 1,500 protesters attending a People's Assembly in Trafalgar Square all agreed that Extinction Rebellion should be inclusive of everyone. The message in my news channel gave a helpful definition of inclusivity. It meant 'reach out to people of colour, the lower-income and the working class, that we should focus our actions against the powerful and not disrupt for the sake of it, and importantly that we keep going bigger and louder, despite bans and attempts to silence us'. 'This is what democracy looks like', the message ended with a flourish.

But the DLR action went ahead anyway. I woke on the morning of 17 October to a brief discussion on BBC Radio 4's *Today* programme, with the interviewer asking why Extinction Rebellion was disrupting public transport and creating problems for a cleaner who had taken two hours to get to work. James O'Brien tweeted: 'I only started coming to work by tube because of Greta Thunberg. And now I run the risk of the train being hijacked by the lunatic fringe of Extinction Rebellion. Would they rather we all drove? Targeting commuters on public transport is a staggeringly stupid move.'[17]

Meanwhile at Canning Town some commuters had responded with violence, pulling a protester off the top of the train. Then other commuters moved to protect the protesters, as the police had taken their time to intervene. All my chat sites were bursting with discussion: the majority unhappy and disapproving, although it was hard not to be moved by the eighty-year-old man who had glued himself to the door of the DLR, stating in a short video that he was getting arrested for the sake of his grand-children. I just wished he had chosen a different target.

Extinction Rebellion takes pride in being a horizontal move-ment to which any individual or group may belong, provided they support the movement's three demands, which are, in brief: 'Governments must tell the truth by declaring a climate and eco-logical emergency; act now to halt biodiversity loss and reduce greenhouse gas emissions to net zero by 2025 [and] create and be led by the decisions of a Citizens' Assembly on climate and eco-logical justice.' Supporters should also commit to underlying principles and values, including 'using nonviolent strategy and tactics as the most effective way to bring about change' and 'autonomy and decentralisation'.[18]

The problem is that if you have a decentralised movement of autonomous groups with no leaders, the autonomous groups will take autonomous decisions and make their own minds up as to what constitutes nonviolence. And the Canning Town activists had some support. An article in the *New Statesman*, quoting BuzzFeed, argued:

> This approach certainly pisses a lot of people off at times but the goal is not to win majority support but to get a large enough minor-ity who would make the cost of doing nothing on climate change higher than taking action to address it. Creating a minority of com-mitted radicals is seen as a surer route to forcing the actions needed on climate change than a sympathetic and concerned but distracted and apathetic majority.[19]

I did not agree. It seemed obvious to me that getting that large enough minority still required winning hearts and minds and not alienating the otherwise sympathetic. Long-time peace activist and researcher Michael Randle was secretary of the Direct Action Committee, formed in 1960 to supplement CND's more conventional protests against nuclear weapons with civil disobedience. In an essay on the role of civil disobedience in a democracy, he argued that:

> Civil disobedience was theatre, was 'propaganda by deed'. For many the nonviolent discipline, the cheerful acceptance of hardship, the willingness to face fines and imprisonment, were seen as ways of communicating a sense of seriousness and urgency of the issues at stake. When successful nonviolent direct action could highlight the issues, touch people's imagination and perhaps persuade them to reconsider their position; it could lead some to take part in the civil disobedience, others to join the larger movement. Thus, although there was a coercive—or at least an obstructive element in civil disobedience, its success depended ultimately upon winning over the population, not forcing views upon it or making it physically impossible for the government to continue its policies by sheer weight of numbers.[20]

I also thought that if disruption was the sole aim, rather than a necessary component of action, personally committed radicals could lose. And when their nonviolent direct action did not produce the results they wished for, they would adopt other tactics to achieve their goal. Some people were arguing that Canning Town made the media pay full attention, but violence always gets covered. My question was, and still remains, how to act radically while causing minimum harm to others. I posted my own thinking on nonviolent civil disobedience online later that day:

1) It is nonviolent, so although I may get hurt myself and be treated with violence by others, or provoked in other ways, I will avoid responding with physical or verbal aggression and do my utmost to avoid physical and psychological harm to others.

2) It is civil: it recognises the legitimate rights of all those affected, including the police, to express their views to me and to protest in the same manner. One can have dialogue through action.

3) It is disobedient: the law breaking ideally is directed at unjust laws. Occupying Trafalgar Square to challenge attempts to limit the right to peaceful protest, and improper use of Section 14, is a good example. Sit-ins at segregated lunch counters, freedom rides on segregated buses are other classic examples. Or we break minor laws to disrupt and draw attention to a greater injustice. Which laws are chosen is a matter of judgment. I want the targets and locations of action to cause minimum harm to those whom we want to support, and who will be most affected by climate change. This group surely includes some of the poorest and most vulnerable people in the City, who live in the Shadwell and Dockland area, are ignored and displaced by global finance and will be among the first victims of any flooding in London. The objects of the action need to clearly symbolise what we oppose and what we support. Temporarily blocking planes from flying and cars from driving causes inconvenience, but symbolises a change we want to see: less traffic and flights. Stopping light rail and tube trains, even those carrying global financiers, suggests we have not worked out our public transport policy.

4) Nonviolence also means the ability to respect and listen to others with whom I disagree. I welcome responses to all the points I have made.

The argument has continued. For some like Rupert Read, a professor of philosophy from the University of East Anglia, Canning Town was a 'crucial turning point' although he had not realised it at the time. It was also the beginning of a slow divorce. I met with Rupert, and learnt about this, sitting among the daffodils in appropriately named Green Park in London in March 2022. Read, a Quaker, has been involved in both anti-nuclear and environmental campaigns since the 1990s. One of his achievements was a small act of rebellion when he refused to debate on the

BBC with a climate-change sceptic. This had resulted in the BBC changing policy in 2018, and no longer balancing scientific discussions on the climate crisis with climate deniers. Shortly after this, he saw Extinction Rebellion co-founder Gail Bradbrook's 'Heading for Extinction' talk on YouTube and met up with her. He joined what was then a small group, became one of their spokespersons and helped to launch the first two rebellions in autumn 2018 and spring 2019. Along with others, he met with Environment Secretary Michael Gove to discuss Extinction Rebellion demands. He told me of his feelings of both disbelief and delight as parliament subsequently declared a climate emergency, held a form of Citizens' Assembly in spring 2020, and passed the law committing them to net zero by 2050. He realised he was part of something 'big in environmental activism that actually changed the whole conversation and made a permanent mark'.

But Canning Town exposed what Rupert regarded as a weakness in Extinction Rebellion's bottom-up form of horizontal organisation. This meant that any group, as long as it subscribed to the overarching principles of nonviolence, could take an action in the way it saw fit, and 'we're in the crazy situation, on the day of Canning Town, that the whole Extinction Rebellion media machine was mobilising to support what had been done, even though hardly anyone in XR supported it'. In an interview with *The Times* Read said that 'personally [he] was sorry that the action had gone ahead'. He got supportive feedback from people via emails and social media, who said they were glad that 'someone prominent in Extinction Rebellion is saying we don't back this, this is not great. This is not the way to go.' But others within the movement accused him of 'throwing brave rebels under the bus', to which he responded:

> I wasn't throwing anyone under a bus—that's a violent metaphor to
> use. And actually, if you are going to use a metaphor like that, then

actually it's the movement that's being thrown under the bus by this tiny minority, which as it were, hijacked not only the name and the reputation, but the actual organisational machinery—and I came to think you can't have an organisation which is going to be long-term successful that is run this way.[21]

I had seen a YouTube discussion of him debating tactics with Roger Hallam. Both appeared to agree that, as Read put it succinctly, 'this civilisation is finished'. Hallam argued as before that a small disruptive minority, say two or three thousand people, engaged in civil disobedience, could force the government to engage in legislative change and that you only needed a small proportion to get that 'rapid flipping effect'. He again used the example of the civil rights movement in the US in 1960s changing the national consensus. Read argued that for that to occur, you needed a 'supportive underbelly', like the mass of the iceberg below the waterline, supporting the visible radical protesters. This could not be created top-down by 'imposing the kind of change that is necessary on the population at large. We have to bring most people with us, otherwise this will not happen, it's a civilisation-level transformation.'[22]

How to bring people with us was something I had been thinking about more and more. On one of those days leafleting at Liverpool Street in 2019, I had been working down the line of stopped cars with a Canadian colleague when a Somali woman driving a large black SUV pulled down her window and screamed at him:

'Are you American?'

'I am Canadian but I live here', he replied.

'How did you get here? You flew didn't you! You flew. Go back home! You want to stop me and my family flying home on holidays, I won't have it!'

She drove off with an angry glare. What we were trying to do was radically different from those earlier movements for the abolition of slavery, or the rights of women to vote or ending racial injustice. We were not asking for rights and freedoms, that were already enjoyed by some, to be extended to all. On the contrary, we appeared to be asking people to stop doing things they enjoyed and to give up consuming stuff they liked; indeed, to change a whole way of life. And the argument that, as our way of life in the Global North was unsustainable anyway, the choice was either letting it be disrupted in a radical and terrifying way by floods or fire, or in a less radical way by getting our governments to help us adapt now, was not an easy one to get across.

Rupert's view was that one way to get across this gulf was to build a 'moderate flank'.

Getting a better understanding of what he meant by this was one of the reasons I had wanted to meet him. Rupert used the term 'moderate flank' for those who, in part thanks to Extinction Rebellion's successful consciousness raising, recognised there was a crisis and craved meaningful action but not 'activism' as represented by Extinction Rebellion. What he wanted and what he believed was already happening was for people to take transformative action within their lives, organisations, workplaces and businesses. In October 2021 he had written an online essay explaining:

> The 'moderate flank' involves a more direct, positive direct action [which] could similarly be read as more radical, in a way. Moving directly to improve business/workplaces (and using strikes/stoppages probably mainly as a resort only if this process is resisted). Moving directly to make communities resilient (and using traditional NVDA [nonviolent direct action] only if this is resisted). [...] This goes beyond 'activism' altogether: and a good thing too, because the term activism is itself non-inclusive. The moderate flank movement needs to permeate and transform everyday life. In my experience, people are now hungry for this positive dimension. They are less interested in protest, more in making stuff actually happen.[23]

'It actually isn't that moderate [...] it's actually still very demanding,' he explained to me, sitting in the park. 'What it boils down to is, right now is your time better spent hanging out with your affinity group and plunking your ass in a road or whatever and feeling very comfortable and right on and pure, or leading or creating a potentially much larger and more significant, moderate flank organisation?'[24]

Others, however, especially after the two years of quiescence imposed by the Covid-19 pandemic, thought that blocking more roads was exactly what was required. In July 2021, a small group of activists calling themselves Insulate Britain came down to Cornwall and held a meeting on a sunny afternoon in St Mary's Church in Penzance. I was one of only very few people attending, but undaunted, the activists explained why they were there. Extinction Rebellion had lost its way and was not doing enough. According to Sir David King, the government's former chief scientific officer, those tipping points of polar ice cap meltdown, and the Amazon rainforest becoming a carbon emitter rather than a carbon sink, had almost been reached. 'So, what we do in the next two to three years will determine the future of humanity', a young man said. What they wanted was to get the government to pass legislation to fund and organise the insulation of all social housing in Britain by 2025, and to produce a legally binding plan as to how all housing stock would be retrofitted by 2030.

'Insulating is one of the easiest ways of decarbonising the economy and it makes sense to people,' the speaker continued, 'and it will reduce energy bills and create thousands of jobs.'

The way they were going to achieve this was not through mass disruption but economic disruption, caused by targeting national infrastructure, specifically the motorway system of South East Britain. They planned to sit down at key junctions and stop the traffic. They hoped the people in the church would join them.

'I like the name and the demands,' I said to the group, 'but I don't quite understand how sitting on motorways and stopping the traffic will help people understand the need for mass insulation of homes? And I can imagine people trying to get home from work getting quite stressed and stroppy.'

'It doesn't matter,' was the response. 'The inconvenience caused will push the government to respond. Look what the Freedom Riders achieved by sitting in the whites-only seats on buses. We are putting ourselves on the line in the same way.'

'Yes, but they were enacting what they wanted to achieve, an integrated public transport system. How do blocked motorways symbolise and help people understand the need for home insulation?' I asked. He had no answer.

Rupert Read made a similar argument in an essay written that September calling for nonviolent direct action that enacted in a positive way what the group wanted to achieve: 'Now that climate-consciousness has been successfully raised, such negative, blunt-instrument actions are simply not going to resonate', he wrote. He suggested instead that Insulate Britain could take Robin Hood-type actions, such as stealing insulation material and distributing it to those with lofts living in social housing. 'Imagine the Government moving to prosecute those who undertook such a programme; there would be howls of disapproval.'[25] What Read was asking for was not less civil disobedience but a change in the language of the action: propaganda by deed.

The motorway blocking actions went ahead. Beginning in September 2021, groups of up to thirty to forty protesters blocked major highways across England. These included the M25 motorway itself, the Port of Dover, roads in Birmingham and Manchester and numerous sites in London, including the Blackwall Tunnel. The public response was, as predicted, hostile. Videos circulated on mainstream and social media of angry motorists berating seated protesters in high-vis jackets and dragging them out of the road. In one, a man walks down the line of seated protesters squirting ink in their faces.[26]

'We're not the people you should be targeting. Do your protest with your placards outside government like everyone else, why stop us from going to work?' An angry man asks on a news video.

'We are really not targeting you', a protester replies.

'You are, you are targeting us! Why stop us from going to work?'[27]

And then there was the incident filmed outside the Blackwall Tunnel where a woman in tears begged protesters to move so that she could follow her 81-year-old mother who had gone to hospital in an ambulance:

> We all believe in what you're doing but I just need to get to my mum. Everybody agrees with you. [...] She's in the ambulance, she's going to the hospital in Canterbury, do you think I'm stupid? [...] I need to go to the hospital, please let me pass. This isn't OK. How can you be so selfish?[28]

Following this incident, Hallam, in a widely reported podcast, said that he would have stayed put had he been confronted with the tearful woman. When he was asked: 'If it were an ambulance and there was someone in there that could potentially die, would you stay there?' Hallam replied 'Yep', and went on to argue that the obstruction was justified given the damage caused by the climate emergency: 'We're talking about the biggest crime in human history imposed by the rich against the global poor.'[29]

Many Insulate Britain activists did not agree with Hallam, Rowan Tilly told me. Rowan is a fellow Greenham woman and a Buddhist, who after a lifetime of nonviolent activism is now a nonviolence trainer and adviser to a number of movements including Extinction Rebellion and Insulate Britain. She told me that she had had to think deeply about blocking motorways, but in the end had decided to join Insulate Britain, because 'putting my hand up in front of the cars was saying "Stop! that's enough now!" Not just those motorists, but the whole shebang, the whole mystique of cars and the carbon economy, all of it, has to stop. That's enough of the destroying this planet and we've got to stop this now.'

But she still felt very uncomfortable when she considered its immediate impact on people. She disagreed with the view commonly expressed by some fellow activists that upsetting people 'did not matter' because the climate and ecological emergency is more important:

> We have to step up to the fact that we are stopping people going about their day, and stopping their work and that this does matter. We can't say that doesn't matter, it does matter, every single person matters, but I still felt it was right to sit on the motorway. [...] I recognised it was an antinomy, an ethical dilemma that can't be resolved. You're still implicated if you walk away.

She told me of one occasion when a doctor who had just had to deal with a suicide and wished to get home to recover, had come up and explained her situation, begging to get through:

> She was in tears and I started crying as well. I said that I understood and I felt awful about it but there's nothing we could do. We had to stay there and this is what it's going to take for us to stop the whole bloody carbon machine. And she looked at me, and she said, very quietly, 'You really mean this, don't you?' And in that moment, I could see that something had dropped in her and she totally got what we were doing. And she just went back to her car. And it was really painful, but I don't regret that we stopped her.[30]

In her paper exploring the ethical dilemma posed by the motorway blockades, she acknowledges the risk of collateral damage, giving the example of causing a car crash, because a motorist in a slowed queue might drive into another car, but she points out that this event could have other causes, such as the driver failing to leave enough space. Therefore one has a duty both to acknowledge this risk, and balance it against the risks and consequences that come from failing to act when people are already suffering and dying because of the climate and ecological emergency: 'If we fail to take action that has a reasonable chance of applying sufficient pressure to achieve our demand on the government then we are neglecting to step up to our civil and ethical duty.'

She draws a parallel with those involved in breaking the Enigma code in the Second World War, who remained 'silent in their knowledge that some ships were to be targeted in order to avoid alerting the Nazis that they had cracked the code and thus be able to issue warning of many more ships to be targeted. So they "allowed" some deaths in order to preserve the lives of many.'[31]

This is the argument for the unfortunate loss of civilian lives during war: the lives of some may be lost in order to save the lives of many. And although I am not a pacifist in all circumstances, I told Rowan that when it comes to trying to achieve political change in a country like Britain, I fall into what she describes as the 'purist' group, who choose nonviolent action because 'means determine ends', and for whom blocking motorways crosses the line of 'do no harm'. My other reason, as I hope to show throughout this book, is that there are other tactics available.

In early February 2022, Insulate Britain announced that they were stopping the road-blocking campaign because 'We have failed [...] to move our irresponsible government to take meaningful action to prevent thousands of us from dying in our cold homes during the energy price crisis. [...] And we failed in getting enough of you to join us on the roads to hold this treasonous and corrupt government to account.' They went on to say that they planned an even more ambitious campaign of civil resistance, ending in a somewhat accusatory tone that seemed unlikely to bring more on board: 'More of you need to join us. We don't get to be bystanders. We either act against evil or we participate in it.'[32]

A few weeks later, someone forwarded me a tweet with a video of a young man in spectacles with longish hair, running onto a football pitch at Goodison Park football ground while Everton were playing Newcastle. Louis McKechnie used a plastic cable tie to attach himself to a goalpost, thus temporarily halting the game. He stared stoically into the distance as the security guard untied him. His bright orange T-shirt said 'Just Stop Oil'. It was just one of a handful of similar actions that month. BBC Newsbeat provided a helpful non-judgmental explainer of the actions and the new group, who were using 'non-violent civil resistance to make their point in public spaces'. The article acknowledged that some people thought the group too extreme and disruptive, but then went on to quote Gary Lineker's tweet that followed McKechnie's protest at Goodison: 'Whether you approve of this young man's methods or not, he's right, his future is perilous. Desperate times and all that.' The BBC article also provided helpful links to 'A really simple guide to climate change'; 'Seven ways to curb climate change'; and 'Small changes you can make for a greener life'; along with further explainers on Extinction Rebellion and Insulate Britain.[33] All that was missing was a simple guide on how to get the government to take action on climate change now.

A week later, the Just Stop Oil coalition blockaded oil terminals across the country by sitting in access roads and climbing on oil tankers. They demanded that the UK government stop expanding oil and gas production because 'it's funding war and killing people in the global South, while destroying the future for young people everywhere'. They called on 'all those outraged at the prospect of climate collapse and suffering from the cost-of-living crisis to stand with us'.[34]

Directly targeting the object of protest, the oil and gas industry, made more sense to me than sitting on motorways to get the government to insulate Britain. Their first blockades had the immediate effect of forcing ExxonMobil to temporarily suspend operations at some of its sites. And I could not ignore the courage and commitment of these predominantly young people who felt they were fighting for their lives.

The dialogue was happening. On ITV's *Good Morning Britain*, celebrity presenter Richard Madeley told young activist Miranda Whelehan that he thought the Just Stop Oil name was childish and playgroundish, and talked about the clothes she was wearing, while she reminded him that 'the latest IPCC report has said we are on the road for climate catastrophe'. The comparison with the movie *Don't Look Up*, in which two scientists try to get across the threat of planetary extinction posed by a meteor heading towards earth, while their breakfast show hosts joke dismissively, was obvious. A video making just that point got 1.5 million hits on its first day online.[35]

* * *

Forty years after I first took direct action in the City of London, my own view remains that we need both disruption and dialogue. We need every kind of action. Nonviolent actions that directly target the oil and gas industry do risk disrupting the lives of some who might not get the fuel they need that day. But they also hold up literally and figuratively the workings of an industry that fuels conflicts like the war in Ukraine, makes record profits while the majority are plunged into poverty, and kills us every day through air pollution, while destroying the possibility of sustaining a liveable planet.

But we also need forms of action that communicate directly with the public, and help us all to engage with this issue. After the Canning Town event, my affinity group in West Cornwall decided to do its own train action. We went out to meet passengers disembarking from the London train at Penzance station to thank them for taking the train, and to give each of them a leaflet about how much carbon they had saved and packets of wildflower seeds.

Asmamaw and I joined Doctors for Extinction Rebellion in 2019. We met them quite by chance during the October rebellion, when they laid out 110 pairs of shoes on the steps in Trafalgar Square to highlight the 110 extra unnecessary deaths caused by air pollution every day in the UK alone, and they invited us both to speak about the climate crisis in Ethiopia.

The movement had been founded by London-based GP Dr Chris Newman. He had always regarded himself as fairly passive and 'centrist' politically, but in 2016, inspired by David Attenborough documentaries, he had started making videos and a podcast about plastic pollution. He was shocked when one interviewee told him that plastic was a much smaller problem than climate change. He decided to attend Extinction Rebellion's first protest in October 2018. However, he stood to the side, helping to hand out leaflets, while others blocked the road, because 'I had this slight aversion. They weren't my kind of people, with the dyed hair, the piercings, the shouting. I felt uncomfortable. Were these just career protesters latching onto another movement? But they were prepared to be arrested ... so I thought I should at least read more about it ... What if they're right?'

The turning point was hearing Sky News anchor Adam Boulton saying to an Extinction Rebellion activist, live on air, that they were 'incompetent, middle-class, self-indulgent people that want to tell us how to live our lives'. Chris found himself wanting to jump into the video to tell Boulton 'Don't you get it? It's not just young people. It's not just hippies. It's professionals too.' And, as Chris explained to me:

> I thought if that was a doctor, or a nurse, or a professor or teacher in that interview chair, someone who wasn't like a person you can easily 'other', and put in a little box saying, 'I don't need to listen to you', he wouldn't have had the balls to be so rude and condescend-

ing. In that moment, a realisation dawned on me. If a group of health professionals supported Extinction Rebellion, it could help legitimise them in the eyes of the public. It could make the difference between this movement living or dying.

He stayed up all night making a website, a YouTube video and social media accounts, and in the morning he messaged everyone he knew. Within days he was sitting in a flat in Dalston with twenty health professionals he had never met before, discussing how they could take the movement forward. Forming Doctors for Extinction Rebellion was itself an act of communication, a statement that Extinction Rebellion was a broad-based movement and did not fit the media stereotype.[36]

For Asmamaw and myself it provided a space where we could actually act upon the issues that disturbed us most: the health consequences of the climate and ecological crises for those most directly affected, in the Global South and at home. So, along with other health workers from all over the country, I have spent hours through the pandemic and beyond on Zoom calls, discussing and designing actions that are visually powerful, and both communicate and disrupt, but overall Do No Harm. Words matter, and however disruptive the action, we have to avoid harming others and we have to communicate and explain what we are doing, whether on a banner, in a leaflet, press release or open letter.

Our actions have included putting 20,000 health warning labels on fuel pumps across the country to alert people to the health damage caused by fossil fuels, inspired by similar labels that are mandatory in parts of the US and in Sweden.[37] In June 2021, when the G7 leaders held a global summit in St Ives, healthcare workers and Extinction Rebellion activists posed as corpses and filled the harbour beach. We held an inquest to dramatise our demands that world leaders should acknowledge the underlying reasons for global pandemics and accept that the climate and ecological emergency is a public health crisis. While some colleagues conducted the inquest, I and others put death certificates on the shrouded bodies. Each stated:

1(a) cause of death: Sars23
1(b) other diseases leading to a: destruction of the natural world
1(c) other diseases leading to b: climate crisis

2 Other diseases contributing to the death, but not related to the disease: air pollution.

And as I placed the piece of card, and the crowds around the harbour watched and listened, I found myself weeping at all that we had lost, and would still lose if we could not turn this around.

And so we have continued both the dialogue and the disruption. In August 2021, more colleagues posed as corpses outside the offices of J. P. Morgan in Canary Wharf, under a banner stating 'Cause of Death: Fossil Fuel Finance', while some of my colleagues in hospital scrubs glued themselves between planted hedges to partially obstruct the entrance. Others lay in the square in front of the tube exit, holding signs saying 'J. P. Morgan: World's No. 1 Investor in Fossil fuels'. I handed out the leaflets I had helped to write, explaining how their money was spent, what fossil fuels did to your health, and the numerous benefits of stopping them, hence the need to stop financing new fossil fuel infrastructure. It was a very different experience to 2019. The visual power of the action caught people's attention and people came up to me to ask for leaflets and to thank me. A tall Black security guard from a nearby office stood reading one beside me, then turned and said: 'You're right!'

A man in glasses and a checked shirt with a trim beard came over. He did not want a leaflet but he did want to chat. He was an investment banker in a neighbouring firm.

'We are trying to educate our customers about this. But they don't want to lose money.'

I WISH THERE WAS SOMEWHERE WE COULD MEET

'Hasn't Mark Carney warned us about fossil fuels being stranded assets?' I asked. He sighed.

'Do you have children?' I continued. 'What do they think?'

'You know, I ride an e-bike and if I am invited somewhere that has nowhere to park, I won't go there. I have two children. One is furious, the other terrified. And all our interns upstairs, you have got them talking. They don't like protesters throwing paint. But this is different. This is an act of communication.'

BREAKING BARRIERS, OCTOBER 1983[38]

'I do basically agree with your principles, I'll admit that to anyone. I don't want to be blown up by a bomb, but we have a job to do.' The young, fair-haired constable sitting beside me in a police van in Newbury Police Station car park had two hours previously arrested me for possessing a pair of bolt cutters and causing criminal damage to the fence at Greenham Common Airbase.

It had, compared to some women's experience, been quite a nice arrest. I had been cutting for at least 20 minutes, while two MOD police, glum and perplexed looking, held firmly onto the fence on the other side—torn between pushing at it to shake me off and holding fast to prevent other women from shaking it down. Wire cutting is very tiring if you are as unpractised as I am, and my arms felt more and more like cotton wool. So, when the officer, playing it by the book, took my arm and said, 'I am arresting you for possessing a pair of bolt croppers, mumble mumble … said will be taken down in evidence …' in a rather hurried mutter, I was quite glad to stop; though sad when another officer deftly fielded my cutters before another woman could pick them up.

Further along, a red-bearded sergeant had brought Jane to the ground in a rugby tackle, bruising and stunning her so that she got into the van loudly and quite justifiably complaining. The sergeant was apologetic, however, and, by the time three more women had been collected, Jane had, in between waving supporters away and telling them to 'pull it down, pull it down', restored everyone's good humour by a rendering of 'Women won the struggle for Greenham when the fence came tumbling down.'

'The thing is,' Jane pointed out firmly, 'there is going to be more and more of this. You are going to like the job less and less.'

Now, sitting in the car park, sharing one officer's cigarettes, we were discussing cruise missiles and the police force. It was clear from the conversation that the officers didn't much like the idea of the missiles. I asked mine why he joined the force—at 20, he had been in it for four years. 'I thought it was worthwhile, you're sorting out problems, solving them.' 'By arresting us?' I asked. 'Well, that is helping others. You're breaking the law and, from where I'm sitting, I have got to.' 'But how do bad laws change if we don't challenge them? How do you think slavery was abolished? That was legal once.' 'I wouldn't have been a policeman then', he replied.

The conversation drifted on to what did they think about the Waldorf case? They felt sorry about it, but thought that the officers were doing their job. 'Mind you, I don't personally want to be armed,' one said. 'I'd seriously reconsider my position if we were.'[39] And back to the bomb. 'I can understand what you are doing today, but I don't think it can be stopped. People at the top make decisions and there is nothing we can do about it. We have all got to die some day.'

More and more vans were arriving, packed with women. Whoops and shouting filled the air. 'At least 55 women arrested so far, it's bloody marvellous,' said Jane. 'What I don't understand is how you could live in conditions like that,' another officer said. 'The thing about Greenham,' Jane explained, 'in spite of all the difficulties, is the compensation in friendship and love.' 'That is exactly why I joined the force', he replied.

At this point, two other officers took over. Tired looking and less good humoured, one complained bitterly that he had been on since seven a.m. and that his wife was heavily pregnant. 'If it wasn't for you ladies, I wouldn't have to be doing this. Nuclear weapons have been here for years. It's bloody rubbish. You know you are wasting your time.' Jane was outraged. 'Wasting my time! I've a nice house in Derbyshire. I've worked all my life to retire to it and you tell me I know I'm wasting my time.' 'Oh dear,' said Vicky, 'we've just been through all this.'

There is no doubt that endlessly trying to explain what we are doing is exhausting; but, for me, establishing that communication is half the point of allowing myself to be arrested. There is a common misconception that Greenham Women engage in acts of civil disobedience simply for the sake of publicity. That is not the case. Civil disobedience, as I understand it, is necessary action that may involve me in breaking the law—either because I need to demonstrate that the law is wrong and should be changed, or because obedience to conscience, or some higher law, will involve the infringement of minor laws not necessarily evil in themselves. Draft resistance is an example of the former; cutting the fence an example of the latter—and one which I believe to be as necessary as breaking down the door of a neighbour's house to put a fire out. It demonstrates that the base cannot operate without heavily protecting itself from the people whom it is supposed to protect.

Also, by taking down the fence we are brought face to face with our 'protectors'. They have to deal with us rather than avoid us, and come to terms with the fact that they have the same choices as ourselves. They could choose not to arrest us. If we are arrested, then we can go on making these choices quite clear.

This is not always easy when you are being brutally treated or dragged through barbed wire, as some women were. Some women therefore choose to communicate their views in other ways, by total non-cooperation; by silence; by symbols. I have always opted for talking, believing that, as new information and ideas have changed me, so they can change anyone if they are presented in a non-threatening form. The difficulty as a woman talking to male police officers is not to let your conversation be misconstrued as flirting.

So, in my van, we talked. We talked from five until eight in the evening, when they shut the five of us into a small, dirty, green cell smelling of feet. The corridor rapidly took on the appearance of a second-hand shoe store.

I waited to be charged with some thirty other women, in a cell designed for ten. Women were crowded onto the benches, lay on the floor, and sprawled across each other. It was very hot. But everyone was good humoured, exchanging accounts of the action and singing songs. Meanwhile, a battered copy of *The Daily Telegraph* circulated. From it, I discovered that Paul Johnson,[40] along with my arresting officer, knows that we all have to die. CND is apparently sweeping death under the carpet by discussing atomic weapons. Curious; I should have thought that it was our very admission of mortality that is so frightening. Taking down fences and exposing the horror on our doorstep is not a struggle against the natural cycle of death and regeneration, but against living evil, against the tyranny of selfish, utterly destructive indifference to anything but one's own material wellbeing, in the here and now.

My own material comfort at that moment consisted of tomato soup and frozen bread. 'I'd have sent someone out,' the Sergeant remarked, 'But 30 more of you have been arrested.' Messages, shouted through the ventilators, told us that four and a half miles of fence was down. At midnight, I was charged, standing formally in a small office. Did I have anything to say? Nothing. I was released on bail for a court appearance the following week. Other women followed. At the base, the fence was still coming down.

3

WHEN DO WE NEED TO BREAK DOWN THE DOOR?

In the autumn of 1988, I received a letter from Air Chief Marshal Sir Michael Armitage, head of the Royal College of Defence Studies, inviting me to give a lecture the following spring on 'Protest—An Alternative View', and to make 'the case for extra-parliamentary action and civil disobedience as a legitimate protest against a democratically elected government'. It would fit into the 'Protest and Conflict' section of the college's annual year-long course. Other lectures in that series included 'The Subversion of Democratic Society', 'Public Order in a Free Society' and 'International Terrorism'.

I leapt at the chance, of course, arriving at Seaford House in Belgrave Square, smartly dressed in my doctor clothes, modestly made up, clutching my notes and plastic sheets for the overhead projector. I was then 'drummed' into the lecture hall where the assembled audience, mostly officers from various armed forces and police, rose to their feet. I had an hour to explain why non-violent activists were not terrorists, and how I believed they enhanced rather than subverted democracy.

SORRY FOR THE INCONVENIENCE

I began by explaining how most of the liberties that we take for granted in Britain, including the most basic rights to hold public meetings and open-air demonstrations, exist because of nonviolent protest, much of it illegal at the time. I asked them to put themselves in St Peter's Fields, Manchester, on 16 August 1819, when between 60,000 and 100,000 men, women and children gathered peaceably to demand constitutional reform. These were the days when only 3 to 4 per cent of the population had the right to vote and the growing city of Manchester had no representative in parliament.[1] Henry Hunt, a leading radical, had asked them to come 'armed with no weapon but that of self-approving conscience, determined not to suffer yourselves to be irritated or excited by any means whatsoever to commit any breach of public peace'.[2]

It was to be a display of the organised power of the unrepresented, and the injunctions of 'cleanliness, sobriety, order' and 'peace' were constantly on the organisers' lips. Hunt had even offered himself up for arrest before the meeting to prevent the magistrates having an excuse for breaking it up. Indeed, like today's generation of nonviolent protesters, those intending to come had drilled in advance to maintain nonviolent discipline. In the words of reformer Samuel Bamford, 'Order in our movements was obtained by drilling; and peace, on our parts, was secured by a prohibition of all weapons of offence or defence.'[3]

By midday there was an enormous crowd, including women and children. Banners waved peacefully calling for 'Universal Suffrage, Election by Ballot, Liberty, Fraternity and Parliaments Annual'. There were no signs of riot. Indeed, 'the peaceable demeanour of so many thousand unemployed men is not natural', one General Byng remarked. It was this discipline and lack of violence that now appeared particularly threatening to those in charge. They realised they were no longer dealing with a rabble, easily contained. This was a visible manifestation of a nonviolent working-class mass movement, making legitimate demands and challenging their moral authority. As E. P. Thompson put it, 'old Corruption faced the alternatives of meeting the reformers with repression or concession'.[4]

They chose repression. At 1 o'clock, the Manchester magistrates decided the whole gathering had the appearance of insurrection and ordered mounted Yeomanry and Hussars to arrest Hunt and disperse the crowd. They charged into the packed space wielding sabres, killing at least eighteen people, including a child, and wounding between 400 and 700.[5]

What followed was a now familiar combination of responses. First, outrage: the public and newspapers of the day condemned both the magistrates and military and there were mass meetings across the country; second, further mobilisation as radical groups sprang up everywhere (including those who called for a more violent response); and third, repression. In December 1819, parliament passed the notorious Six Acts. These, among other things, forbade meetings of more than fifty people, allowed for unauthorised searches of private homes for weapons, toughened the laws for seditious and libellous writing, including the penalty of transportation, and increased the stamp duty on periodicals, making them too expensive for most people.

Nevertheless, the tide had turned. Peterloo, as the massacre had been labelled in ironic reference to the Battle of Waterloo in 1815, helped establish the right to public meetings and contributed greatly to a moral consensus that parliamentary reform was necessary. In 1832 the Reform Act was passed, providing among other things, representation for at least some of the people of Manchester, although it would take another century of pressure to achieve universal suffrage.[6]

I went on to show how often governments, when challenged by effective nonviolent protests that are legal, still find ways to make them illegal. For example, 'combining' to form a trade union was no longer illegal in 1833, when Methodist minister George Loveless helped agricultural workers in the village of

Tolpuddle in Dorset to form the Friendly Society of Agricultural Labourers, in order to protest against their drop in wages and demand 10 shillings a week. So the government, increasingly alarmed by growing agricultural unrest after two years of poor harvests, arrested them for the 'unlawful taking of oaths', for which they were tried and sentenced to twelve years' transportation to Australia in 1834.[7]

In 1955, when Rosa Parks was arrested for refusing to give up her seat to a white person in Montgomery, Alabama, the subsequent bus boycott organised by Black citizens was not illegal. But the city authorities and police then sought to counter the effectiveness of the boycott by declaring that carpooling, arranged to help Black people get to work, was illegal.[8]

My audience had no difficulty accepting that events like these contributed to the rights we enjoy today. What they found more difficult to swallow was the idea that law-breaking was in any way justified in a modern democracy that provided the right to peaceful protest. Indeed, a quick audience poll showed that 99 per cent of those present thought law-breaking in any circumstances was wrong. I then asked how many in the room had ever parked on a double yellow line: 75 per cent put up their hand. 'So, breaking the law for your own convenience is OK? But not in order to challenge an unjust law or an unjust society?'

They laughed. What I wanted to get across to an audience engaged in law enforcement and security, any of whom could be called in by civil authorities to deal with protesters like myself, was that laws could on occasion be unjust and illegitimate, and that democratically elected governments could act illegally. I shared with them the position of moral philosopher John Rawls, who argued that:

> In a viable democratic regime there is a common conception of justice by reference to which its citizens regulate their political affairs and interpret the constitution. Civil disobedience is a public act

which the dissenter believes to be justified by this conception of justice and for this reason it may be understood as addressing the sense of justice of the majority in order to urge reconsideration of the measures protested and to warn that, in the sincere opinion of the dissenters, the conditions of social cooperation are not being honoured. For the principles of justice express precisely such conditions, and the persistent and deliberate violation in regard to basic liberties over any extended period of time, cuts the ties of community and invites either submission or forceful resistance. By engaging in civil disobedience, a minority leads the majority to consider whether it wants to have its acts taken in this way, or whether, in the view of the common sense of justice, it wishes to acknowledge the claims of the minority.

In the face of injustice, civil disobedience offers an alternative to both passivity and violence, thus Rawls regarded it as a 'stabilising device in a constitutional regime tending to make it more just'.[9]

The law can be wrong. I pointed out that the British *Manual of Military Law* (1958) states that 'Obedience to the order of a government or a superior, whether military or civil, or to a national law or regulation, affords no defence to a charge of committing a war crime.'[10] Thus it appears to accept that legitimate governments might on occasion demand illegitimate action. I and my fellow Greenham women friends had argued that the possession and planned use of nuclear weapons broke both British and international laws. Britain, for example, was a signatory to Article 51 of the 1977 Protocol Additional to the Geneva Convention, which states clearly that combatants may only direct their operations against military targets and must spare civilian populations and objects.[11]

As Hiroshima and Nagasaki had clearly demonstrated, nuclear weapons do not make this distinction and thus must be illegal. Moreover, the UK Genocide Act of 1969 defined genocide as acts that had 'the intent to destroy, in whole or in part, a national, ethnical, racial or religious group'. These acts included 'killing, causing serious bodily or mental harm or deliberately inflicting on the group conditions of life calculated to bring about its physical destruction in whole or in part'.[12] Nuclear weapons certainly do all of these things. So, what were our choices? Doing nothing? We knew where that path led. But violence was unacceptable. Unlike terrorists, as nonviolent activists using civil disobedience, we were not trying to force change. Rather, we wanted to hold up a particular government action for public examination—the undemocratic and illegitimate deployment of nuclear weapons, that threatened genocidal violence—and persuade both the public and government to change their minds. And by the time I gave this lecture in the spring of 1989, presidents Mikhail Gorbachev and Ronald Reagan had signed the Intermediate-Range Nuclear Forces (INF) Treaty, committing both the United States and the Soviet Union to the complete removal and permanent elimination of their intermediate and shorter-range land-based missiles, both nuclear and conventional. So, at least regarding cruise missiles, we had succeeded.

After taking a break, we came back for an hour of discussion and this was where the crunch came. Supposing one accepted that nonviolent civil disobedience had a legitimate place in a democracy, the audience wanted me to clarify how I defined nonviolence. What were its limits? Was damaging property nonviolent? How could I justify cutting the fence at Greenham?

It was the same point my grandmother had made in a brief letter written to wish me well when I went to Newbury Magistrates' Court in March 1984 for the fence-cutting action. 'I cannot accept that violence will help your cause', she wrote. I wrote back at once:

> I cut down a fence in the same way as I might amputate a leg if someone had gangrene that might kill them. Removing the fence that protects the most violent weapons ever designed, might encourage

them to take those weapons away. I have never, nor will I ever use violence against any person. [...] Knowing me and knowing how much I care about pain and suffering I hope you would trust me when I say I will have nothing to do with anything that causes it. What do you think of police breaking down a door to put out a fire?

I had then used this defence of necessity in the courtroom, arguing that, as allowed under Section 5, parts 2 and 3 of the Criminal Damage Act 1971, I had lawful excuse for causing damage, because it was my honest belief that property and life were in danger, and that the means I was taking to protect them were reasonable. To my surprise, Stipendiary Magistrate Barr did not prevent me explaining in detail how cruise missile deployment increased the likelihood of a pre-emptive nuclear strike that would damage not just Newbury and everyone in it, but much of the surrounding country. He listened patiently as I catalogued once again what flash, blast and radiation do to the human body. Then I summed up:

> I pulled down the fence with others as the only means available to me to actually prevent a damaging agent being placed in a situation where it could endanger lives and property by turning them into targets. It was a protective action, analogous to destroying timber to prevent the advance of a forest fire and done to prevent the onset of damage.

I also pointed out that this defence did not require him to share my beliefs, or regard them as justified, or to share my view that my actions were reasonable. The law allowed me to be mistaken: 'If you accept that I honestly believed that I was taking reasonable action to protect property from immediate danger, then you must find me not guilty.'

Mr Barr disagreed. He found me guilty and fined me £50. I refused to pay and so was sent to prison for two weeks.

The Plowshares Eight in the United States went much further. In 1980, eight Catholic activists, including Daniel and Philip

Berrigan, brothers who had previously burnt their draft cards to protest the Vietnam war, broke into a General Electric plant in King of Prussia, Pennsylvania. Once inside, they hammered on the nose cones of the nuclear warheads and poured their own blood on documents. They took the biblical call from the prophets Isaiah and Micah, to 'beat swords into ploughshares', literally. Molly Rush wrote about that action forty years later:

> I banged my small household hammer on one of the warheads and made a small hole in it. A tiny piece hit me under my chin. I was able to put dents in the other one. I thought, they're not really invulnerable. They're really not. We're vulnerable, but so are they. And that means that human beings have made these, and human beings can unmake them. The others hammered away, then poured baby bottles of blood on them. We then circled and prayed and waited to be arrested.[13]

However, at their trial the judge refused to hear either expert witnesses or the justification defence, and they were sentenced to prison terms of between one and ten years. After a decade of appeals, they were finally sentenced to time served. The action inspired a global Ploughshares movement which continues to this day. Small groups of individuals have taken direct actions that involve damage to military hardware or property. What they have in common is that while they attempt to directly damage warfighting equipment, no harm is ever done to a person. In addition, the activists take personal responsibility for their actions and do not flee the scene. Almost all have been prosecuted and found guilty, often facing sentences of between one and five years in jail. In the United States in 1984, two individuals who had entered a Minuteman missile silo and damaged the silo cover were sentenced to eighteen years.[14]

There was one notable Ploughshares acquittal. In 1996 three women were arrested for criminal damage after damaging the electronics of a Hawk jet at the British Aerospace factory in

Warton, Lancashire. They were part of a Ploughshares group called 'Seeds of Hope'. After causing more than £1.7 million worth of damage and dancing around the hangar, they still had to phone the press to get security to actually arrest them. Their justification was that they had to prevent the plane reaching its buyer, the Indonesian government, who would use it against the East Timorese people. A week later, a fourth woman was arrested and charged with conspiracy. All four pleaded not guilty, 'on the basis that we had lawful excuse as we were acting to prevent British Aerospace and the British Government from aiding and abetting genocide', as one of their number, Joanna Wilson, explained.[15] The jury at Liverpool Crown Court agreed, and they were acquitted.[16]

Is this kind of damage to property nonviolent? I think so. No one was hurt or endangered and all of these actions were clearly done to prevent violence being inflicted on human beings. Andreas Marcou, a researcher on civil disobedience and the law, argues that damage to property can be considered to be nonviolent civil disobedience when it passes what he calls the safeguarding and communication tests. This means that the actions must '(a) safeguard the bodily and psychic integrity of other agents, avoiding acting in ways that intend to cause or are reckless about causing harm to other people and (b) embody political communication'.[17]

There are more ambiguous cases. Does setting fire to heavy machinery, using oxyacetylene blowtorches to cut through exposed pipelines, and setting fire to valve sites in a number of secret acts of sabotage against the controversial Dakota Access Pipeline in the United States in 2016, count as nonviolent? Jessica Reznicek and Ruby Montoya, members of the Catholic Worker Movement, argued that it did, seeing the risk to millions from water contamination caused by oil leaks and spills as a greater threat. Having tried petitions, marches, hunger strikes, boycotts and encampments without effect, they felt they had no choice. They recognised that their actions were controversial, but insisted that they were both necessary and 'never threatened human life nor personal property'.[18]

In 2021 and 2022, Reznicek and Montoya were given prison sentences of eight and six years respectively, partly because of the application of a terrorism sentencing enhancement, and because Judge Goodgame Ebinger said the harsh sentences were necessary to deter others. The prosecution argued that because the defendants did not know what was in the pipes when they used the blow torches, they put first responders and construction workers at risk.[19] The question has also been raised as to whether the women were manipulated into taking this particular kind of action by government operatives who had infiltrated the protest movement.[20]

Interestingly, not all the actions of the suffragettes would pass Marcou's standards. I had learnt at school that their struggle to obtain the vote included window-breaking actions in Oxford Street, the slashing of the Rokeby Venus, setting fire to post boxes and, most famously, Emily Davison sacrificing her own life by throwing herself in front of the King's horse in 1913. I also knew of their courage in facing the violence inflicted upon them, particularly the force-feeding that was used to stop their hunger strikes in jail. I had included them in my list of exemplars of civil disobedience in my lectures at Seaford House. Extinction Rebellion activists today frequently cite them as a source of inspiration.

It was only recently that I discovered how this campaign of damage to property and self-sacrifice was deliberately escalated to include arson and bombing attacks. Beginning in 1912, the most intense period was between early 1913 up to the summer of 1914, after a franchise bill, that would have included amendments giving women the right to vote, was dropped in the House of Commons.

WHEN DO WE NEED TO BREAK DOWN THE DOOR?

Partly inspired by the Irish Republican bombing campaign of the 1880s, some suffragettes, the majority of whom were paid employees of the Women's Social and Political Union (WSPU), set fires and planted homemade bombs in private homes, outside public buildings and in sports grounds, theatres and churches. They also targeted public infrastructure such as railways, canals and telegraph wires.[21]

Lives were definitely put at risk; some people were killed and others injured. For example, Mary Leigh threw a hatchet at Prime Minister Herbert Asquith when he was visiting Dublin in July 1912. It missed, but cut the ear of his companion, Irish MP John Redmond. Later that day she and three other suffragettes tried to blow up the Theatre Royal in Dublin, as people were leaving a packed performance. They placed a cannister of gunpowder near the stage, poured and set light to oil under seats, which filled the dress circle with smoke, while Gladys Evans was caught in the process of throwing lit matches into the cinematographer's box full of flammable film reels. If the fires had not been discovered and put out, many might have been killed.[22] On another occasion, bombs were discovered at the South Eastern District Post Office in London containing enough nitroglycerine to blow up the entire building and kill the 200 people who worked there. Fortunately, no one was hurt.[23]

But others were. A number of postmen were burnt by the chemicals suffragettes used to cause postbox fire. On 13 July 1914, *The Manchester Guardian* reported a train guard having his hands and arms badly burnt when a mail bag exploded, also setting other bags and the train carriage alight. More postal workers were severely burnt in similar incidents in Nottingham.[24] The historian C. J. Bearman estimates that between 1913 and 1914 alone there were some twenty to thirty-five incidents when life was threatened. And lives were lost. Two men were killed by a fire started in a naval dockyard in Portsmouth in 1913; and it's possible that both men and horses died in fires in Bradford that were claimed by *The Suffragette*, the newspaper published by the WSPU.[25]

These are just a handful of instances from a campaign of militancy that the WSPU leadership did not deny involved violence. Indeed, the arson and bombing attacks were detailed in weekly double-page spreads in *The Suffragette* under the headline 'Reign of Terror'.[26] Christabel Pankhurst had no scruples in acknowledging the violence implicit in these acts, writing in 1913:

> If men use explosives and bombs for their own purpose, they call it war, and that the throwing of a bomb that destroys other people is then described as a glorious and heroic deed. Why should a woman not make use of the same weapons as men? It is not only war we have declared; we are fighting a revolution.[27]

Her mother, Emmeline Pankhurst, was similarly clear in one of her most famous speeches, stating:

> It is clear to the meanest intelligence that if you have not got the vote, you must either submit to laws just or unjust, administration just or unjust, or the time inevitably comes when you will revolt against that injustice and use violent means to put an end to it.[28]

And they recognised that people would suffer, writing in the WSPU's Seventh Annual Report of 1913:

> That private citizens should be affected is inevitable, for this is war, and in all wars, it is the private citizen who suffers the most. It is, in fact, by means of pressure on the private citizen that an opposing force finally achieves its victory. Moreover, in the women's war for the Vote, the private citizen cannot complain of suffering the pains and penalties of warfare, because there is nobody who can plead innocence and irresponsibility where the question of votes for women is concerned.[29]

Which is to say, the public deserved what they were getting, because they were complicit. Reading these accounts, I think it very likely that both Emmeline and Christabel would themselves have rejected the descriptor of nonviolent civil disobedience. But the violence alienated their own supporters within the already fragmented Suffrage movement. Emmeline Pethick-Lawrence was expelled from the WSPU for opposing the violence in 1912. Christabel's sister Sylvia left to form the East London Federation of the Suffragettes in 1913. But both Emmeline and Christabel Pankhurst believed that this violent form of militancy was needed because it would terrorise and coerce both the public and government into conceding to their demand for votes for women. Arguably, it had the reverse effect. In May 1913, the last time women's enfranchisement was debated before the outbreak of the First World War, Liberal MP Harold Cawley said: 'These out-breaks [of suffragette militancy] are carefully calculated, stage managed, cold-blooded crimes committed by a very few members of the public and directed by a few well-paid and highly advertised leaders. It may be right to yield to the violence of the many, but I am perfectly certain that it is bad policy to yield to the violence of the very few.'[30] Prominent suffragette supporters such as Labour MP Philip Snowden agreed, noting in a speech in 1914 that 'The women's actions during the past year had so set the clock back that the suffrage question is temporarily as dead as Queen Anne.'[31] Arguably, it was the role women played during the First World War that would earn partial suffrage for those over the age of thirty in 1918.

I have gone into this in some detail because a century later, defining what we mean by nonviolence matters. Extinction Rebellion has an absolute commitment to nonviolence, including not blaming and shaming, yet many activists give the suffragettes as an example, as I did in the past. And the argument that the public are complicit and deserve what they are getting is not dissimilar to the argument used by some climate activists regarding the inconvenience caused by blocking motorways. This is not to suggest the damage done by blockades is equivalent to that caused by planted explosives, but we need to know and understand the history we cite.

Extinction Rebellion also have a video on their website that asks: 'Can Property Damage be Nonviolent?' In this, former Seeds of Hope activists Jo Blackman (formerly known as Joanna Wilson) and Rowan Tilly, who had been part of the support group for the action against the Hawk jet back in 1996, explain the care they took to ensure the action was not intimidating or threatening to anyone, did not risk harm or injury, and that they were accountable. They spent some months in preparation, making sure that the action 'would not place anyone at physical risk' through, for example, an explosion, but they would disarm and disable the plane to ensure that it could not be used as a weapon. Witness statements and a video about why they were taking the action were deliberately left at the scene of the crime. The video included statements that had been made in parliament about what was happening in East Timor, and a commercial from British Aerospace saying their planes delivered 'a powerful punch'. They wanted to expose the hypocrisy and criminality of the British government selling arms to the Indonesian government, which had been condemned by the UN for its illegal, genocidal occupation. As Jo Blackman explains in the video: 'When the police talk about the scene of the crime they're talking about the criminals [which] was supposed to be us, but actually what we did was we turned that around completely and said, no, it was the British government and the arms trade that are the real criminals here.' She recommended it as a strategy to be used by other activists: 'I'm not sure why we haven't seen more memory sticks with evidence files left at actions, because when something is left as evidence at the scene of the crime, it has to be shown to the jury.'[32]

Many Extinction Rebellion actions in the last few years have involved damage in the form of activists glueing themselves to buildings and/or painting, chalking or pasting on buildings. Angela Ditchfield used the same necessity defence that I had tried in front of Mr Barr, thirty-five years previously, when she appeared in Cambridge Magistrates' Court in November 2019, charged with criminal damage for spray-painting the headquarters of Cambridgeshire County Council the previous year. She claimed that she had lawful excuse because she had tried all legal means to persuade her local council to take action on the climate crisis, through meetings, letters and petitions. When the council's business plan for 2018–19 appeared without a single reference to climate, she felt that she was left with no choice: 'Contributing to climate change as we all do, amounts to criminal damage to the whole web of life on earth, and violence against its poorest people.' The lay bench in the Magistrates' Court agreed, and I was jealous when I read their verdict:

> We find that you have a very strong and honestly held belief that we are facing a climate emergency, and that you acted on the spur of moment to protect land and homes under threat from climate change, believing that immediate protection was necessary, and the action could be said to have been taken to protect property, and that you believed the action chosen was reasonable in all circumstances ... We find you not guilty.[33]

Also in 2019, a jury found Roger Hallam and David Durant not guilty of £7,000 worth of criminal damage after they argued that chalking 'Divest from oil and gas' on the walls of King's College London in 2017, was a proportionate response to the climate crisis.[34] A jury came to a similar conclusion in April 2021 regarding six activists who had broken windows, poured fake oil, hung a banner above the entrance and sprayed messages on the wall of the Shell headquarters in London in April 2019.

Interestingly, Judge Perrins allowed the Shell defendants to put their arguments as to why it was their honest, subjective belief that their actions were necessary. He did then instruct the jury that this defence of necessity had no legal basis, because of a previous High Court judgment which said that 'the defendants' acts must be considered in the context of a functioning state ... in which even the most passionately held beliefs can be expressed in a way which does not involve committing a criminal offence'. Judge Perrins's implication was that in a functioning democracy you don't need civil disobedience, let alone criminal damage. But the Shell defendants argued that:

> in this area of the climate emergency, it was our honest belief that we did not live in a functioning state at all. We were allowed to provide evidence to support that belief: evidence of successive governments' failures to act; evidence of the success of big oil in lobbying against regulation; and evidence of the deaths of thousands of UK citizens from the effects of air pollution as a result.[35]

One of the defendants, David Lambert, closed his defence by arguing that there was a moral necessity for the jury to find them not guilty:

> Our case is not strong in law but we feel it is strong in conscience: we would not be here if we had not acted on the basis of our conviction that we must do whatever it takes to make government recognise the emergency for what it is—not just a phrase for politicians' speeches, but a barely imaginable horror, no longer on some distant horizon, but unfolding in real time in the real and beautiful world all around us.

In his summing up, Judge Perrins made his views clear: 'This is a court of law, it is not a court of morals. I have given you clear direction on what the law is and your duty to apply that law to the facts as you find them to be.' He went on to remind them of the oath they had taken to 'reach true verdicts according to the evidence', adding, 'those are not mere empty words. A true verdict is one that is reached having all due regard to the law. That is how our jury system works and that is what you all pledged to do.'[36] The twelve members of the jury took just over seven hours to find all the defendants Not Guilty.[37]

Damage does appear to be leading to dialogue. In March 2021, Extinction Rebellion co-founder Gail Bradbrook took a hammer to the windows of her local Barclays Bank in Stroud to draw attention to their role as the biggest European funders of fossil fuels.[38] A few weeks later, seven women carefully broke the glass at Barclays headquarters in Canary Wharf. They placed stickers on the glass saying 'In case of climate emergency break glass', and wore patches that read 'Better broken windows than broken promises'—referencing the glass-breaking of the suffragettes. They then sat down and waited to be arrested.[39]

A year later, tech and finance commentator Chris Skinner invited Bradbrook to make a monthly contribution to his daily blog, Finanse.com. This was met with fury by some of his regular readers. Skinner quoted one former bank chief executive on his blog, who wrote:

> You cannot have activist groups who are breaching civil rules taken seriously. Extinction Rebellion have broken my bank's windows in both branches and head office and encouraged protests that disrupt the economy, such as access to Heathrow airport. For me, this makes them an extremist organisation that you should not be encouraging to take part in our community.[40]

In her next blog, Bradbrook responded:

> Those of us who undertake acts of peaceful civil disobedience do so with a clear intention to be accountable for our actions. We take great care that no one is harmed and wait for the police to come and arrest us. [...] These are not 'mindless acts', they are deeply caring acts by thoughtful and courageous people. They are also not necessarily illegal. We do not intend to break the law, so much as challenge the status quo; one which is destroying life on earth.

She went on to point out how an earlier act of criminal damage had led to her being invited into a number of other prominent dialogues. She had taken a hammer to a window at the Department of Transport in October 2019, in protest at the environmental impact of the high speed railway project, HS2. The BBC had then invited Bradbrook to ask the opening question to Dr Mark Carney, former Governor of the Bank of England, now UN Special Envoy for Climate Action and Finance, following his fourth Reith Lecture in 2020 on the climate crisis. Carney had agreed that social movements had shifted people's attitudes and resulted in better understanding, a political consensus and legislative change. Meanwhile, a former vice chair of a major high-street bank had told her, in confidence, that 'provided no harm is done to individuals (including the harm from blaming and shaming), then as far as I'm concerned there is going to be a need for this kind of action because change has not come quickly enough'. He thought bank employees agreed with her, but the leadership could not act because they were trapped 'in systems skewed in the wrong direction'. That's why he supported 'extreme' views and actions as long as they were peaceful.[41]

I would love to know his opinion on throwing two cans of tomato soup at a Vincent van Gogh painting of a vase of sunflowers. In October 2022, Asmamaw and I were doing voluntary work in Guatemala when my Dutch neighbour came over to ask me what I thought of the two young women who had just done this in the National Gallery in London, before glueing themselves to the gallery wall. I told him that I supported the activists. Just a few days earlier, Tropical Storm Julia had caused devastating floods and landslides across Guatemala, destroying roads, homes and schools and killing at least eight people. Apparently, the glass-covered painting was completely undamaged. Would he and I be discussing the climate crisis without that action? I watched videos of 21-year-old Phoebe Plummer speaking directly to camera, asking:

> Are you more concerned about the protection of a painting or the protection of our planet and people? The cost of living crisis is part of the cost of oil crisis, fuel is unaffordable to millions of cold, hungry families. They can't even afford to heat a tin of soup. Meanwhile, crops are failing. Millions of people are dying in monsoons, wildfires and severe drought. We cannot afford new oil and gas.[42]

They were both arrested for criminal damage and aggravated trespass. The outrage on social media echoed across the planet as well. A day later, Plummer was standing next to a can of Heinz soup in an orange jacket explaining herself on a TikTok video that had gone viral, getting more than 8 million views. She assured us she would not have carried out the action if there was any chance of damaging the picture:

> We're not asking the question should everybody be throwing soup on paintings. What we're doing is getting the conversation going so we can ask the questions that matter [...] I'm stood here today as a queer woman and the reason that I'm able to vote, the reason I'm able to go to university, hopefully someday marry the person I love is because of people who have taken part in civil resistance before me.[43]

As to what Van Gogh himself might have said, many of the comment pieces quoted him: 'It is not the language of painters but the language of nature which one should listen to, the feeling for the things themselves, for reality is more important than the feeling for pictures.'[44]

* * *

Diary: 7 June 2020, Newlyn, Cornwall

I am lying flat on my back on Newlyn Green next to Asmamaw. Both of us wear homemade masks saying 'I can't breathe'. Around us are other members of our affinity group, carefully socially distanced, and some sixty other adults and children. Some take the knee, some stand, almost all are holding placards: Black Lives Matter, Racism Kills, Our Silence Will Cost More Lives. Mine quotes Martin Luther King: 'There is no such thing as an innocent bystander.' At 11 o'clock we all fall silent for nine minutes and as I lie there, I realise what a long time that is, and how horrifying it would be to take that time to choke to death. The image of the policeman Derek Chauvin kneeling on George Floyd's neck in Minneapolis, that I first saw on Twitter, floats into my mind. His expression as he stares at the photographer is flat and indifferent. It seems to say, 'So what? You think this is problem? This happens every day.' His colleagues do nothing to intervene as Floyd chokes to death. Again, rage surges within me.

We had to do something. What's going on in the United States just now feels like civil war. In response to the largely nonviolent Black Lives Matter protests across the US, Trump has authorised the largest military deployment ever in peacetime. On Twitter I have watched videos of police cars driving into crowds in New York City; journalists thrown to the ground; demonstrators hit unprovoked with batons. In a particularly horrifying one, an unarmed elderly man is knocked down by two police officers in Buffalo. He is left unconscious and bleeding on the ground as the police walk by, and one actually restrains another officer who wants to help![45] *And then there is Trump's walk to the church of St John's in Washington, with police tear-gassing peaceful protesters out of the way to make his path safe. Something has shifted. It is as if all the horror and misery, the racism, lies and hypocrisy, half buried under the surface for years, is now explicit. The United States and the United Kingdom, who built their wealth and power on a criminal trade so many years ago, are now world leaders in the mortality statistics from Covid: and guess who is most vulnerable? The Black and ethnic minorities whose daily work helps all of us to stay alive. Failed states both of them.*

Asmamaw holds my hand and I stare up as the grey clouds above me coalesce into butterfly shapes. I can hear the sea, voices far off, the quiet murmur of a child. The nine minutes are up and we get to our feet. The world continues as it was, but it means something to put your hand to the lever to try and tilt it a little.

Others are tilting it even further. After getting back from the protest, I watch the BBC News covering the hundreds of thousands demonstrating in all the major cities in the UK and all over the world. One action in particular catches my attention. In Bristol, thousands of young protesters have pulled down the statue of the slave trader-cum-'philanthropist', Edward Colston. I watch amazed as a dancing, cheering crowd of predominantly young people throw a rope around the statue's neck, and topple him off his plinth. One man kneels on Colston's neck. Some jump on the statue; then four of them roll the body, with his paint-covered face and his rope-bound ankles, down to Bristol Harbour, where he is thrown into the dock.

It's an astonishing act of political communication. I look up Colston. In the seventeenth century he shipped more than 84,000 Africans into slavery, including 12,000 children.[46] *He did this while Deputy Governor of the Royal African Company, which, according to the historian David Olusoga, transported more Africans into slavery than any other company in the whole history of the slave trade in the North Atlantic. The Company branded its enslaved captives with the initials 'RAC' and then imprisoned them in the hellish slave ships that plied the Middle Passage. It is estimated that 19,000 died on those voyages and were thrown to the hungry sharks who had learnt to follow these ships. The survivors were sold to work tobacco and sugar plantations in the Caribbean, and the profits and goods from that trade were brought back to Bristol Harbour.*[47] *Now his effigy has been thrown back into the waters from which those ships sailed and returned.*

No one was hurt during the action, it seems, and in contrast to the United States, the local police force have tried to avoid inflaming the situation.[48] *But already there is an outcry. Accusations of mob violence, vandalism and attempting to erase history are flying round social media. 'Utterly disgraceful', Home Secretary Priti Patel declared this evening.*[49]

SORRY FOR THE INCONVENIENCE

Perhaps the disgrace is that there is a statue celebrating Colston as 'one of the most virtuous and wise sons of their city' while many in Bristol have been campaigning for years for its removal.

Far from history being erased, George Floyd's death, combined with this singular action, appears to be stimulating a major awakening and debate on the role of slavery, colonialism and racism in society today. Suddenly we are all re-examining the monuments, paintings and street names around us, the hymns we sing, and the history we were taught in school, thinking about what it means and who and what is being remembered and celebrated.

* * *

My own ignorance is a case in point. In history classes at school we looked at the horrifying engravings of slaves tightly packed into the hold of the *Brookes* slave ship and learnt about Britain's role in the abolition of the slave trade. But William Gladstone, for example, was presented as an early defender of human rights, decrying Ottoman brutality in the Balkans. No one told us that his father was one of the largest slave owners in the West Indies, or that Gladstone supported the payment of £20 million in compensation to slave owners, as well as the apprenticeship system that tied emancipated adult slaves to their former owners for a number of years. Nor did he want his father's compensation payment of £93,000 for 2,000 slaves made public. Writing in 1877, he explained that the sufferings of the Bulgarians and Slavs at the hands of the Turks had moved him more than the sufferings of Black slaves at the hands of their white masters because 'in the case of negro slavery [...] it was the case of a race of higher capacities ruling over a race of lower capacities'.[50]

In English class we read and admired novels by Charles Dickens, but never discussed his racism. This was a man who wrote to a friend after the 1857 Indian rebellion that 'I should do my utmost to exterminate the Race upon whom the stain of the late cruelties rested [...], proceeding, with all convenient dispatch and merciful swiftness of execution, to blot it out of mankind and raze it off the face of the Earth.'[51] Does that mean I want to stop reading Dickens? Absolutely not, I just want a fuller understanding of who he was.

I went to medical school in Bristol in the 1970s and lived for some years in rented rooms in a house in St Pauls. I walked past Colston's statue almost every day with little thought as to who he was, or what he had done to have schools and the concert hall named after him. I took my exams in the Wills Memorial Building, named for the University of Bristol's first Chancellor, whose wealth from his tobacco farms, worked by slaves, made the founding of the university possible. And while I joined marches by the Anti-Nazi League to challenge the growing power of the National Front, I had no real understanding of how many of my Afro-Caribbean neighbours suffered from poor housing and education, or discriminatory treatment by the police. These miseries erupted into riot in 1980, and forty years later Black people constituted 24 per cent of homeless households in Bristol, despite only making up 6 per cent of the total Bristol population.[52]

Discussions about statues had been going on for some years, but the Colston toppling gave them new life. In June 2020 there were also mass street demonstrations in front of Oriel College, Oxford, where British imperialist Cecil John Rhodes stared down unamused from above the door. The Peace Pledge Union, whose members had booed the Queen Mother when she unveiled a statue to Arthur 'Bomber' Harris outside St Clement Danes Church in London in 1992, renewed their demand that the statue of Harris should be removed. This was, after all, a man who had developed and advocated the bombing of civilian populations as a war-fighting strategy. He began in the Middle East, while stationed in Egypt during the Palestinian Arab Revolt in 1936, suggesting 'one 250-pound or 500-pound bomb in each village that speaks out of turn'. The strategy was refined during the last three years of the Second World War. In 1945, 25,000 people were killed in a single night in Dresden, because Harris believed that the war could be won by the destruction of German cities, the killing of German workers and the disruption of civilised life throughout Germany. Geoff Tibbs, the Remembrance Project Manager of the Peace Pledge Union, thought the atrocities committed by Bomber Harris in Iraq, Germany and elsewhere were not something to celebrate. He pointed out that the RAF were building on the legacy of 'aerial policing' by training Saudi troops, who were bombing schools and hospitals in Yemen at the present time.[53]

Books and documentaries discussing these issues have long been available, but it was only after the Colston action that I started to really think about and explore them. That singular and dramatic nonviolent action of pulling Colston off his pedestal was not about erasing history, to quote Olusoga again, but 'making it'. It was truly propaganda by deed, and by shining a light into dark holes that had been closed or forgotten, it made us ask ourselves who we wanted to remember and celebrate, and how.

It also reminded me and other environmental activists that we have to acknowledge, discuss and address the manner in which the long shadow cast by racism underpins climate injustice. Both at home and across the Global South, the main victims of the climate and ecological catastrophes are people of colour, many from the same populations whose enslavement and transportation across the world has helped us build the industrial, carbon-based economies that are killing them right now. And still we have not repaid what is owed.

So do I think all the statues I do not like should be pulled down? Certainly not. The solution the Indian government adopted of moving the statues of its previous imperial overlords to a public park is an interesting one. It's a pity that the plan to create an educational open-air museum around them has never been developed. Actually, I agreed with some of what Boris Johnson said after someone added the words 'Was a Racist' to Churchill's name on his statue in Parliament Square. In a short rant on Twitter, Johnson wrote:

> We cannot now try to edit or censor our past. We cannot pretend to have a different history. The statues in our cities and towns were put up by previous generations. They had different perspectives, different understandings of right and wrong. But those statues teach us about our past, with all its faults. To tear them down would be to lie about our history, and impoverish the education of generations to come.[54]

Which was exactly why I liked the graffiti addition to the statue of Mr Churchill. I think having statues that teach us about our past, with all its faults, is a very good idea. It's not about editing or censoring information, but adding to it. That is exactly what was proposed for Colston. Throwing him in the dock was not the final act. He has since been pulled up and put on display, complete with graffiti, in the M Shed museum in Bristol. A survey commissioned by the council polled 14,000 people, of whom 74 per cent said they wanted him to continue being exhibited lying on his side. A report by the 'We Are Bristol' History Commission called 'The Coston Statue: What Next?' also recommended that the plinth and original plaques should remain in place, and that a new plaque be installed to explain why the statue was first put up and when it was taken down.[55] The process provides a model of what could be done elsewhere.

In December 2020, the four activists involved in toppling Colston were charged with criminal damage. Their trial began in December 2021. William Hughes QC argued for the prosecution that 'what Edward Colston may or may not have done, good or bad, [is] not on trial and [is] not an issue for you—these four defendants are'. In other words, they had no lawful excuse.[56]

The defendants disagreed. They did not deny taking down the statue, but Milo Ponsford argued that he did have lawful excuse because the statue was disgraceful and offensive to local people and he wanted 'to prevent further harm to the people of Bristol'. Sage Willoughby said he had been signing petitions to have the statue removed without effect for decades: 'Imagine having a Hitler statue in front of a Holocaust survivor—I believe they are similar [...] That was not an act of violence, that was an act of love for my fellow man.' The judge permitted David Olusoga to appear as an expert witness, detailing Colston's history and role in the slave trade.[57]

The jury found them not guilty by a majority verdict and all four were acquitted in January 2022.[58] Rhian Graham, another of the Colston Four, wrote in *The Guardian* shortly afterwards: 'I don't think that this verdict means we should start pulling down all the statues in the UK. Really, it's not about statues at all: it's about that statue, in this city, at this time. [...] Our case has demonstrated the value and power of protest.' Moreover, far from damaging the statue, its monetary value had increased fifty-fold since its toppling:

> In that sense, how can it be said that we damaged anything? That statue is a far more useful tool for history and learning than it ever was before, which negates any of the arguments made about us 'erasing history'. You can't erase history. What Colston and the myths around him have done is shroud history by deeming him—as the statue's plaque says—'one of [Bristol's] most virtuous and wise sons'. We are trying to shine a light in places people don't want lights shining.[59]

IN THE EYE OF THE STORM, DECEMBER 1983[60]

'In a way I suppose it's a compliment to us', Woo remarked. We were staring at the rolls of barbed wire lining the perimeter fence of the base and beyond them an inner wall of three rolls of wire. The searchlights placed at intervals had just come on and a Ministry of Defence policeman walked slowly towards us. 'Strange compliment', I replied. And we turned away pulling branches out of soggy undergrowth for campfire fuel.

Life at Greenham is full of those nice ironies that make a journalist's trade easier. I can scramble from my 'bender' in the morning to gasp at the beauty of frost-caught bracken and birch, light glittering on grass, and a pied wagtail poised on a branch above my head, then turn and remember that the reason I am here is the ugliness behind me. I can reduce myself to a feeling of total impotence and rage, standing by the flood-lit, wire-encased silos late at night: yet at the same time, watching soldiers reprimanded for talking to us and eventually pulled right back, wonder what it is I have to say that is so powerful. And while swearing in short-tempered exhaustion at the mud and the rain and the impossibility of finding a clean cup when I brush my teeth in the morning, I never cease to be amazed at the burst of creativity: of poems, pictures, songs and humour that all this seems to produce.

The greatest paradox of all, of course, is that while the press keeps asking us 'Why, now that you've failed, don't you all pack up and go home', more and more women keep arriving and the camp becomes more entrenched. The various components have gained a style and identity that only comes with habit and use: Bender City—the oldest benders built on Ministry of Transport land by injunctioned women; the clearing fire, a suburban sprawl of tents and newer benders put up by women coming down after the October action. There are now six camps, each at a different gate. Women who camped here in the first months are coming back and others who like myself previously only stayed one or two days a week, stay permanently. 'There is a feeling that now you have to be here', Lesley said, one of the early campers.

The vigilantes, however, have quietened down; they have burned no benders, and thrown no shit for at least a month. And it is some time since I have been woken by male voices muttering 'fucking cunts', 'dirty lesbians', as heavy feet tramped outside. The presence of increasing numbers of women and bright lights seems to have frightened them away, while those in Newbury who do support us have made themselves more visible, appearing daily with a banner to stand across the road from the main gate.

These physical changes are important, reflecting an increasingly besieged and defensive attitude on the part of those who support cruise missiles. But there are other aspects of the camp that far from changing have only become more pronounced.

The most obvious of these is the continuing commitment to women-only space and women-only actions. The detailed reasons for this vary, as does, after two years, the patience with which they are expressed. But the consensus remains the same: women-only actions and a women-only space, far from dividing the movement, appear to enable thousands of women to take action who would never have thought of doing so before; moreover, the kind of action that emerges appears to be totally different from what comes out of mixed groups.

IN THE EYE OF THE STORM, 1983

There is no doubt that weaving gates shut with wool, climbing into the base in teddy bear suits, dancing on silos, lighting candles, keening, to name but a few, could all be condemned as 'ludicrous nonsense', as a smartly dressed public school boy told me at a talk at his school. They appear, however, to have been effective in a way that 500,000 gathered in Hyde Park have not. Dressing as witches added a dimension to the so-called 'real' act of cutting down fences that completely changed its nature, making it not just a challenge to cruise, but to the whole set of assumptions on which nuclear defence rests: that power is a justifiable end in itself and that hierarchies based on violence and coercion are suitable forms of social organisation.

The fact that there are now women's peace camps as far apart as Pine Gap, Australia, and Seneca Falls in New York State and that Greenham has inspired women to direct action all over the world would seem to vindicate this argument. The actual arrival of cruise missiles is seen not as a sign of ineffectiveness, but as a clear signal that even more of the same is obviously needed; which is why statements from parts of the peace movement that 'now is the time to change to mixed action' are reviewed with some amazement. 'What do they think we've been doing for two years, if now that it's suddenly got tough we're going to turn round and ask the men in to help!', one woman exclaimed.

How, in fact we do work—even after two years of doing it—I still find difficult to describe. It is easier to say what it is not, as Ceri did under cross-examination during the trial for dancing on the silo in February of this year.

'Who planned the action?' Christopher Fry asked for the prosecution.

'It was nobody's idea, we all meet together, a suggestion comes up and it evolves', Ceri replied.

'Then who suggested it?' he persisted.

'I don't remember.'

'Who was involved?'

'All of us.'

'All 44?'

'No, more; 1,000s are involved, we come from all over the world.' Ceri was not being deliberately obtuse, simply truthful.

In spite of attempts by the media and the authorities to create them, there never have been any leaders. Women clearly identified by the press as spokeswomen have no more impact on decision-making than women who may have arrived 10 minutes ago. And while this may mean that a decision made one week is changed the next, causing endless frustration for more organised movements who want to 'deal with the same person' to 'plan ahead', it has been the camp's main protection. Newbury District Council found to its cost that injunctions and evictions, far from removing the 'hard core', just increased its size.

Now the term 'Greenham Woman' has grown to embrace not just camp women, but any woman who chooses to take non-violent direct action for peace. A loose network of women now spreads around the country, seen most vividly in action on 24 May and again on 9 November, in setting up a hundred and two peace camps. Communications are by word of mouth, phone; messages are at best haphazard. There is no membership, no regular newsletter, conflicting information is often given out about the same event and yet somehow it continues to grow, continues to surprise the authorities, and the women within it become increasingly determined to confront the government on their own terms.

So far, the authorities have refrained from using the heavy penalties used 20 years ago against the Committee of 100. It is unlikely that, if they did, they would at this stage have the necessary deterrent effect. Increasingly arbitrary and discriminatory treatment by the Magistrates' courts—as in the case of Sarah Hipperson being remanded in custody for challenging the Magistrates' bias and three women being imprisoned for two weeks for contempt—has not deterred women from getting arrested. It has simply led us to an increasing sophistication in dealing with the courts.

As to the future sometimes the sense of living in the eye of a storm makes it hard to see clearly where one is going. Events whirl round you, caught briefly in newspaper headlines, then are whipped out of sight before you can catch your breath, and the only calm place seems the spot where you are standing. Greenham in some ways is an easier place to be than anywhere else, in that at least you are right up against it.

That is not to say that on occasion one doesn't feel like a gladiator in a ring. Never more was this so than on 14 November when TV crews descended on us, radios were thrust in our faces and our reaction to Defence Secretary Heseltine's announcement filmed as we listened. None of us has any desire to be that 'brave lunatic woman who gets herself shot' as James Cameron suggests.[61] On the contrary, many of us who, like Paula, arrived because of that announcement are here because we want to 'live to have grandchildren'.

What I have done in a confusing and often depressing year is managed to clarify one thing for myself. That the slogan 'means determine the ends' is not just rhetoric. I no longer believe cruise can be stopped by simply a massive physical presence in one place, even a non-violent one. That way lies endless confrontation. The importance of Greenham for me is in showing that it's going to take a sea change in people's consciousness to turn opinion polls into effective action. Actions that don't simply block the launchers, which we certainly intend to do, but challenge the whole edifice of militarism.

Action that comes from taking personal responsibility, being flexible, creative, spontaneous and able to trust.

We harp on about these alternative ways of working not simply because they have enabled us to continue or because we have lost sight of the main issue—cruise—but because these alternatives seem very much to the point in challenging missiles that depend on a rigid and destructive way of life. Diane, another plaintiff in the silo trial in February, was also extensively questioned by the prosecution:

'Did you go into the base by accident or desire?'

'It was by desire—to celebrate life.'

'How often do you intend to go to Greenham Common?'

'Until the government takes us seriously.'

'Do you expect it to take you seriously?'

'I do. I represent half the world's population.'

4

WHO ARE YOUR LEADERS?

Diary: 7 April 2022, London

This time it's a tall white-haired man, elegantly dressed in a dark overcoat. He walks down the steps of HM Treasury and along Horse Guards Road towards me, eyeing my six colleagues who are glued to three oil drums labelled Gas, Oil and Coal, blocking the street. Asmamaw stands with others at one end, holding a banner saying 'For Health's Sake: Stop Financing Fossil Fuels'. As usual, I am handing out leaflets, and the man walks up to me, reaches out saying 'I would like one', then tears the small piece of card in two and hands it back: 'Here you are!'

'Thank you very much', I reply, as he walks away. It's a pity because it was a good leaflet on which we had spent quite some time, explaining the damage to health caused by fossil fuels, and the health benefits of getting rid of them. I've got a new catch phrase, calling out 'Happy World Health Day' as people pass. It makes most people smile and perhaps because it's a sunny day beside St James's Park, quite a number stop to chat and take one.

We have been here since 7 a.m. The Daily Express has already labelled us an 'Eco Mob'. The police have helpfully closed the road

111

at both ends except to cyclists and emergency vehicles, which we always let through. Earlier, Asmamaw and I delivered a letter to Chancellor Rishi Sunak, from Doctors for Extinction Rebellion, asking him to:

> Remove the financial incentives that the government gives to fossil fuels, which amount to around £10 billion/year according to the 2019 EU Commission report, and use this money instead to support clean energy and energy efficiency strategies, whilst supporting the poorest families out of energy poverty.[1]

The letter had to be taken to the back entrance to be checked for dangerous substances. The actor Laurence Fox turns up to berate us for not being at work saving the NHS, and to berate the police for failing to arrest us. He films all of us for his Facebook page, where he labels us 'eco-loons' and shows film of himself asking the police why they are nicer to us than they are to anti-vaxxers, but he does include me inviting him to join us.[2]

How to communicate across this gulf? Three days earlier the IPCC [Intergovernmental Panel on Climate Change] produced the third part of their Sixth Assessment Report which stated clearly that we had to cut emissions drastically before 2025. According to Jim Skea, one of the co-chairs of the working group behind the report, it was 'now or never, if we want to limit global warming to 1.5°C. Without immediate and deep emissions reductions across all sectors, it will be impossible.'

Two days after that, the UK government produced its new energy strategy committing itself to creating new oil and gas infrastructure, but saying nothing about insulation. This in spite of a poll by the Energy and Climate Intelligence Unit which showed that 84 per cent of those asked thought insulation was an important way to limit government reliance on gas from Russia. Insulate Britain looks prescient.[3]

I spent the train journey to London trying to make sense of the disconnect between the two, and writing a furious letter to my MP:

WHO ARE YOUR LEADERS?

Dear Derek Thomas,

Last week after watching 'Don't Look Up' you told us all to wait, and that you were optimistic about the new energy strategy. Yet this is what we are going to get. More investment in fossil fuel infrastructure and no investment in insulation that would make a real difference to fuel poverty. It looks like the meteor is continuing to hurtle towards us while your colleagues are partying.

I begged him to listen to the video message from UN Secretary-General António Guterres, in which he called the IPCC report:

A file of shame, cataloguing the empty pledges that put us firmly on track towards an unliveable world. [...] We are on a pathway to global warming of more than double the 1.5°C limit. Governments and corporations [...] are adding fuel to the flames. They are choking our planet, based on their vested interests and historic investments in fossil fuels, when cheaper, renewable solutions provide green jobs, energy security and greater price stability. Investing in new fossil fuels infrastructure is moral and economic madness.[4]

10 April 2022, London

Words matter but they are not enough. Three days after our protest at the Treasury, I am now sitting with some thirty medical colleagues dressed in scrubs blocking Lambeth Bridge, as just one small part of a week of disruption organised by Extinction Rebellion. We marched here from Hyde Park along with some few thousand others, and after sitting with a couple of hundred people listening to speeches on an open mike, we were asked to have a quick people's assembly to decide whether we wanted to stay here or join the larger group blocking Vauxhall Bridge. Asmamaw and I sat with our health worker colleagues. Everyone had a chance to speak and give their views. Everyone agreed that we would continue to block this bridge as health workers when the others had left. The idea was to sit and make a positive health-focused protest. No one objected. The decision was taken back to the wider group. They also agreed, and so we walked through the crowd with our banner stating 'For Health's Sake: Stop Financing Fossil Fuels', to sit at the end near Victoria Embankment, as everyone else left for Vauxhall Bridge.

A large number of police officers arrive in high-vis jackets and line up in front of us. A couple of police liaison officers in blue waistcoats walk amongst us informing us that the bridge is now subject to Section 14, so we are likely to be arrested. The traffic on the Embankment is quiet, but when an ambulance wants to come through from the south, everyone immediately jumps to their feet shouting 'blue light' and sprints to the side of the road. We then sit down again. The police superintendent steps forward and addresses us through a loud hailer as 'Dear NHS colleagues'. He confirms that the Section 14 has now been imposed on this bridge, and that if we don't move, we will be arrested. Asmamaw is observing from the pavement. I am sitting with Fiona Godlee, former editor of the British Medical Journal. *We've both decided to move when asked. So, when two young policewomen come up and explain that I will be arrested if I continue to sit there, I get up.*

As I do so one of the policewomen asks me why I am doing this. 'Because in the places where I work people are dying now because of droughts and floods caused by the climate emergency. A total of 17 million people face starvation in East Africa right now. Do you read about that in the papers?' She shakes her head. 'I simply do not know how else to get your attention.' She looks devastated. Seven of my colleagues, including Chris, remain seated until they are arrested, then stand up silent and dignified to be searched and taken off to Charing Cross Police Station in a van.

'Climate activists are sometimes depicted as dangerous radicals,' António Guterres *said in that video message, 'but the truly dangerous radicals are the countries that are increasing the production of fossil fuels.'*[5]

* * *

'Did you know you were going to do that action?' a friend asked me when they saw the pictures. Unlike the Treasury action, which had received little press coverage, the health workers sitting on Lambeth Bridge had appeared on the BBC, and in the *Metro*, *Mail* and *Evening Standard*, as well as the *BMJ*. 'I thought you didn't agree with blocking highways?'

'Early morning on an almost deserted Horse Guards Road, and Saturday afternoon on Lambeth Bridge, after the police have closed it to traffic, are very different from blocking motorways during rush hour. And we always move for emergency vehicles.'

'Who planned it?'

'All of us on the bridge at that time took the decision together', I explained.

'And the Treasury? That must have taken some organising.'

It did. As with the action at J. P. Morgan that preceded it, we held a series of lengthy Zoom discussions with any supporters who wanted to be involved. Someone—I don't remember who—suggested the action in the Doctors for Extinction Rebellion chat group to which supporters belonged. Those interested in taking the action or supporting it then set up a separate chat, and we met regularly online to talk until both the design of the action and the practical details, such as how to bring in the oil drums, writing and printing leaflets, and the time and place of meeting were sorted out.

One of us would volunteer to facilitate and another to take minutes, and decisions were made with a show of hands to check for consent. If everyone agreed, the plan went ahead. If there was a veto, the action would need to go back to the wider group to be discussed.

It is one of the most democratic movements in which I have been involved. There are founders, but no leaders. I asked Chris Newman if he had planned it that way. His answer was no, that it had developed organically with the momentum coming from all those engaged. Early in 2020 I had attended an open meeting in East London, bringing together health workers from many different climate-focused groups. For example, friends and colleagues in mental health had set up a group called 'Divest Psych' to pressure the Royal College of Psychiatrists to divest from fossil fuels, declare a climate emergency, and develop policy on mental health and climate change. There were others working to make their practices or hospital trusts greener and more sustainable. It was an extraordinary day, with a feeling of energy and possibility, and with the pleasure of hanging out with like-minded colleagues.

Then the Covid lockdown came and we all migrated online. But in some ways, setting up chat groups, working in a digital space was even more democratising. It meant that movements like Doctors for Extinction Rebellion and Divest Psych did not work from a physical centre in which those with time and proximity to the office dominated. I could participate in strategy discussions while sitting in Ethiopia, where I was working at the beginning of 2022. A nested system of chat groups has emerged, including an administrative hub, and a general 'all can post' area. Thematic and action groups form and dissolve according to need and interest. This is in addition to the website, Facebook, Instagram and TikTok pages that introduce Doctors for Extinction Rebellion to the wider public and provide news of actions and a means to contact the administrators.

A structure has evolved. Obstetrician Alice Clack explained it to me in December 2022. Doctors for XR has a number of teams or circles that do organisational work, including administration, communications and outreach; as the movement has increased in size, regional and local groups have emerged, each of which send a representative to the regional chat. Doctors for XR supports the demands, principles and values of Extinction Rebellion and communicates with the movement, but organises separately from it.[6]

Indeed, there is an eleventh principle shared at welcoming meetings: this is 'professionalism', 'in recognition of the trust and value our professional status as health professionals holds within society, and an understanding that we have a duty to uphold this while taking action as a health worker group'.[7]

According to Chris and Alice, anyone can join the mailing list and participate provided they can prove their identity as a health worker. And the importance of including all kinds of health workers has grown. Recently, after lengthy discussions and a systematic process of online polling, we voted to change the name to Health for Extinction Rebellion, to better represent those involved. By spring 2023, signed up supporters numbered more than 1,000.

The digital world may have enlarged the democratic space for movement activists, but it has also made it more exhausting. I recently sat down to count the chat groups I belong to. Currently there are at least six main ones, and that does not include the temporary action-related subgroups. The problem is that unlike old-fashioned meetings, you cannot just leave and shut the door; the chatter goes on day and night. Learning to simply turn off the chats is an essential survival skill.

There is 'Psych Declares', the name adopted by Divest Psych after the Royal College of Psychiatrists did divest from fossil fuels and declare a climate and ecological emergency.[8] It appears that conventional pressure from trusted messengers can work, especially within institutions where members share common ethical goals. Which is why I also support Medact, a more formal, membership-based, charitable organisation that grew out of the former Medical Campaign against Nuclear Weapons. It now has a broader agenda and supports health professionals campaigning against the health threats posed by climate change, and structural violence, as well as violent conflict.[9]

Then there is Dolphins, my local Extinction Rebellion affinity group, which is just one of the numerous autonomous local groups across the country that stretch from Orkney to the Channel Islands. This is part of a global movement that includes more than 1,000 Extinction Rebellion groups in eighty-six countries, including many in the Global South.[10] As Helen Angel, our internal coordinator, reminded me, 'anyone who follows the 10 core principles and values can take action in the name of Extinction Rebellion. Using nonviolent civil disobedience to get the government to meet the three demands, we can act autonomously.'[11]

This autonomy means that although I had participated in three national actions before sitting down to write this chapter, I had not had the time or felt the need to explore how Extinction Rebellion organised nationally. I trusted the external coordinator of our group, who connected to the county group, which connected to the forty groups in the South West regional group, whence another external coordinator participated in the central organising circle, or 'Rebel Hive'.

I can view the whole Extinction Rebellion structure, laid out as a mass of overlapping circles, on the 'Self-Organising System' section of the organisation's web page.[12] It looks rather attractive—a map of blue and green bubbles connected by golden lines to larger bubbles. The Central Rebel Hive contains all the regional representatives, as well as connections to groups that have arisen out of shared identity, rather than shared geographical location, such as Health for Extinction Rebellion. It also includes external coordinators from organisational teams such as the operations circle and the strategy circle. Each click also provides a description of the tasks for which that group is accountable.

Extinction Rebellion developed a constitution in 2019, to which all members of Central teams must adhere. All volunteers must also sign a volunteer agreement. 'It is complicated,' Mark Wingfield agreed, but 'no more than a typical hierarchy'. Mark is a jazz musician and sound engineer who got involved with

Extinction Rebellion in early 2019 and is now a coordinator in the Self-Organising System's circle which is responsible for 'maintaining and improving the Extinction Rebellion UK constitution', which he helped to write, and 'supporting the implementation and understanding of our self-organising system. The Extinction Rebellion UK constitution provides rules for self-organising based on holacracy.'

We were talking on Zoom in December 2022. 'There's no such thing as a structureless group,' he told me, 'structure is either formal and visible or informal and invisible.' He was quoting 'The tyranny of structurelessness' by Jo Freeman, a familiar author for any woman activist of my generation, in which she points out that:

> For everyone to have the opportunity to be involved in a given group and to participate in its activities, the structure must be explicit, not implicit. The rules of decision-making must be open and available to everyone, and this can happen only if they are formalized. [...] The more unstructured a movement is, the less control it has over the directions in which it develops and the political actions in which it engages.[13]

Erica Chenoweth argues that successful civil disobedience movements are highly organised and disciplined, and maintain their organisational infrastructure even under pressure. She does acknowledge that leaderless movements are much more difficult to subvert or decapitate. This was certainly the case at Greenham in the early 1980s, where we recognised the likelihood of infiltration by both our own and other security services (as later turned out to be true), but knew that they could not dominate or take control.[14] However, Chenoweth also points out that that leaderless movements are less able to negotiate with opponents, have more difficulty managing public relations crises, building working relationships with other movements, or evaluating and adapting strategy as needed.[15]

'The fact is,' Mark told me, 'if people don't have a shared set of rules there will be misunderstandings and they will bump into each other. If this kind of mass civil disobedience is going to be successful, according to the evidence it needs to be highly disciplined and organised.'

Extinction Rebellion UK now uses its Self-Organising System in a much more disciplined and informed way than in the early days, Mark explained:

> Making difficult decisions by consensus is slow or impossible. Asking everyone to have a say in difficult decisions too often results in polarised views which can't be reconciled. Someone has to make the decisions in a movement where we have many contrasting and strongly held opinions. Mass civil disobedience movements that have not delegated decisions in one way or another, have ground themselves to a halt. Occupy is often given as an example of this.

Extinction Rebellion UK's Self-Organising System delegates responsibility for particular decisions to specific groups (called teams or circles) and individuals, so that when needed, decisions can be made quickly but the decision-maker will be held accountable. So, overall strategy is made by a strategy team elected by the national and regional coordinators. The organisational work needed to implement that strategy is now done by task-oriented teams who have the authority to make decisions in their own areas of responsibility, and who have external coordinators who connect with the wider group.

Marshall Rosenberg, the American psychologist and founder of the Center for Nonviolent Communication, pointed out that the most important decision to be made collaboratively is who makes which decision.[16] This is what the Extinction Rebellion constitution sorts out. The argument is that distributing decision-making in a disciplined and coordinated way to these semi-autonomous teams prevents the chaos and clumsiness that can arise in large, structureless movements. This holacratic structure 'avoids the concentration of power that occurs in patriarchal, hierarchical organisations, and which has contributed to the world's present crisis', Mark told me.

There are other checks and balances. All coordinators must stand for election every six months. There are also circles delegated to deal with conflict and harmful behaviour, as well as those working on inclusivity and accessibility.

'So could an action like Canning Town happen again?' I asked him.

'I think it's pretty unlikely now, but not impossible. While regions and nations have signed up to the constitution, local groups have not. We cannot force it on them and the full set of rules would be too much for a small group.'

'Most action ideas do come from outside the centre,' Mark continued, 'but for an action to be taken up nationally it has to be agreed by the Actions circle.'[17] In addition, the fact that the national movement offers many resources, training, materials and connections to other groups makes its support extremely valuable. When Health for Extinction Rebellion has taken actions like those at J. P. Morgan, or outside the Treasury, it is national Extinction Rebellion volunteers from the arrestee and legal support circle and their media teams, who have helped out. We could not do such actions without them.

Does it matter how movements are structured and organised? I think so. For many, involvement in some form of pressure group or movement is their first experience of political engagement of any kind. If they feel their voice is heard and respected and there is a role they can play, they will be empowered and, as happened with so many Greenham women, they will go on to engage politically in multiple ways. A democratic nonviolent movement provides a positive model of how to act and engage in politics at a time when we are constantly reminded that 'politics is broken'.

* * *

SORRY FOR THE INCONVENIENCE

Diary: Election night, 12 December 2019, Penzance

There is a sudden silence. John just read out the BBC exit poll telling us that Boris Johnson stands to win the biggest Conservative majority in parliament since Thatcher's in 1987, while Labour are about to suffer their worst losses since the 1930s. The silence continues as we all scan websites confirming our worst nightmares. The heaviness in the room is in complete contrast to the laughter and easy banter of the previous couple of hours.

I am sitting in front of a computer at a desk in the Liberal Democrats' party office alongside two other Green Party members who, like myself have spent the last two hours and much of the previous week, canvassing for a party that is not our own. That fact alone says something about the current state of British democracy.

I've discovered I quite like phone canvassing and to my surprise, apparently have a small talent for it. No doubt all those years of listening to patients helps. It is also a lot easier on chilly foggy nights than knocking on doors. The not-at-homes and wrong numbers in the virtual phone bank can be marked up and skipped over in a second, and the immediate hang-ups after I have introduced myself, or the very posh county voices saying 'I am sorry, but I don't want to know, thank you', are less demoralising than a door quickly closed in my face. In fact, very few people hang up. It seems that introducing myself with 'Hello, I am Dr Lynne Jones and I am a Green Party member supporting Andrew George the local Liberal Democrat', arouses sufficient curiosity. Then, as at work, simply asking what issues are bothering them most, and listening, has a good effect.

Even Conservatives seem to enjoy a chat. A Tory farmer from the Lizard [peninsula] wanted my opinion on the rate of caesarean sections at the local hospital; he felt it was too high. Explaining that I was an ignorant child psychiatrist did not stop him wanting to share his experiences inducing lambs.

I am happiest with the undecideds. Then I do my own spiel focusing most on whatever is concerning them. 'I like Andrew George but I want

Corbyn as prime minister' is an easy one, explaining that the best way of helping that to happen is voting tactically to unseat our local Conservative. Unfortunately, the local Labour candidate has no chance of doing that, whereas Andrew was only 300 votes behind in the last election. Wavering Greens are the easiest because I tell them I completely share their views, but I am not voting for my own party because climate change is an emergency, and we cannot afford five more years of fossil fuel subsidies, or a third runway at Heathrow under Tory rule. And if we don't get rid of this government, we have no chance of the electoral reform that would give the Greens a genuine chance of power.

Quite a few of them say 'Thanks, I was thinking along those lines myself', or, even better, 'Yes, I agree, you can count on my vote.' This results in an adrenalin rush, and smiles from my co-workers. It's completely addictive. Perhaps I missed my vocation as a telesales marketeer ... Except I don't think I could be convincing without believing in what I was selling, in this case a local man whose personal positions on the things I care about, the climate and environment, social justice and health, are similar to my own.

Today we've simply focused on encouraging people to go out and vote. They weren't all friendly. Earlier this evening a woman screamed at me down the phone, 'I don't appreciate being rung at this time of night ... and as for the leaflets I have had through my door I could have grown two bloody trees ... and all of you ringing on Sundays and in the evenings, I'm fed up with the whole bloody thing. You are all the same!'

'I am so, so, sorry to disturb you. The only reason for calling, apart from reminding you to vote, is to check if you needed any help to get to the polling station.'

There is a sharp intake of breath: 'Did you know I was disabled?'

'No, but we ask everyone if they need any help.' Slight mollified, the woman thanks me and tells me all her family have voted, and I promise to feed back complaints about leaflets.

The next was a young Cornishman telling me: 'No, I am not voting. It's a waste of time. I done Brexit, nothing happened so what's the point? I work forty-eight hours a day and these politicians got no respect, there's no point and it's a waste of time. Sorry love, but I'm not voting.'

I am sad, but don't try to argue. This morning Asmamaw and I walked through the misty grey and gloomy village to cast our own votes. We took our ballot papers and made our crosses. When Asmamaw and I put our paper votes in the ballot boxes it struck me again how significant it was. I do have some small say in who and how I am governed. Asmamaw reminded me as we walked home, carrying milk from the shop, that this was only the second time he had voted in his life. He never did in Ethiopia because the ruling TPLF's thirty-year control and manipulation of the vote made it meaning-less. The irony is that as Ethiopia now stumbles towards democracy, I feel we teeter on the brink of catastrophe. I vote, but does it count? I wish we could have learnt from Poland, where in some constituen-cies in the last general election, opposition parties worked together, fielding only one candidate. In every place they did that they won, defeating the conservative Law and Justice Party. When I got home, I summoned up the courage to post my video explaining why I am voting tactically for the Lib Dems, and not for the Green Party candidate, on the Cornwall Green Party website. I have had lots of likes on my own Facebook page, but what's the point of making the video if I don't share it with the group that matters?

By 9.30 p.m. I've made more than a hundred calls, at least two-thirds positive—like the elderly lady who says 'Hello my lovely, don't you worry now, you tell that Andrew George I voted for him, I always do.'

But now the warm bubble those calls created has burst.

Friday, 13 December 2019

Johnson's victory is confirmed, and all the chat groups I belong to are gloomy and despondent. I post my own message on all the chats to which I connect:

WHO ARE YOUR LEADERS?

I am old enough to remember 1983 and the feeling of absolute despair at losing that election at the height of the Cold War. I was living outside a missile base and a few months later the missiles arrived. But four more years of nonviolent resistance saw them removed, two years later nonviolence ended the division of Europe, and a year later poll tax protests got rid of Mrs Thatcher. We can change things. Being part of this movement gives me hope. Never give up.

I am astonished at the number of likes and messages from people I do not know, thanking me for being positive. This lifts my spirits somewhat, at least until mid-afternoon when St Ives declares. Conservative Derek Thomas has won, taking half of last year's Labour vote, but if Liberal Democrat, Green and remaining Labour voters had united around a single candidate they would still have beaten him by 237 votes.[18] Already the websites have worked it out: using the proportional representation system that is used in the European elections, we could have had a rainbow coalition running the country.[19] As it stands, the three parties (Lib Dems, Greens and UKIP) who got 16 per cent of the vote between them, gained 2 per cent of the seats in parliament. The Lib Dems actually increased their vote but lost seats. When will we learn?

* * *

That election night was the end of my engagement with party politics. Five years earlier, when we came home from working in Mozambique, I had joined the Green Party, naively believing that party engagement might be a way to bring about change. I thought I would be joining a group that would be out campaigning on climate and environmental issues, but local meetings were completely taken up with how to win the next election. As one member explained to me, 'you have to capture power before you can engage in the issues that matter'. The fact that we lived in a constituency where because of the first-past-the-post system, we had no chance of capturing power, but every chance of handing it to the Tories through dividing the opposition vote, did not seem to matter. This was exactly what had happened in 2015 when the Green Party share of the vote was just enough to deprive Andrew George, then our Liberal Democrat MP, of the few votes he needed to beat a Tory challenge.[20]

When we held a post-election analysis meeting the day after the 2019 election, our local candidate asked angrily why no Green Party members came out to canvas for him. 'Because the majority of us were out working for Andrew,' I answered, 'and if the party had taken an email poll of constituency members rather than taking a decision in a single general meeting, which many could not attend, it might have learnt that most of us did not want you to stand.'

I was chastised for breaking party rules by campaigning against our own candidate. My argument that I said nothing against anyone, merely campaigned FOR someone else was not accepted. I ended with begging everyone to put aside tribal loyalties and to work for a progressive alliance. This got a small round of applause but the committee response was the usual: 'We've tried but Labour won't, so we won't ...'. Back to the children's playground. It was my last contribution. The following June, during lockdown, I had a disciplinary hearing on Zoom in which I was interrogated by a number of men who asked if I was I sorry.

'I am sorry to have broken the rules.'

'Would you do it again?'

'I hope it would not be necessary because we would take a more intelligent approach to the issue of forming a progressive alliance.'

'Did you campaign on the street?'

'Yes, along with half a dozen other Green Party members, every night.'

'Are you planning to join the Liberal Democrats?'

'No, I don't like the party, but what I wanted was to get rid of our Tory MP because of the damage he was doing.'

I also pointed out that my whole strategy was to get the Greens into power by getting rid of Tories now. Shortly after, I received the ruling by email:

> The complaint was upheld. The Disciplinary Committee is satisfied that the Party has been brought into disrepute under section 4.7 of the Constitution. In considering the appropriate sanction, DC felt that the Respondent lacked any remorse or insight into the damage her actions had caused to the Party. The Disciplinary Committee ruled that the Respondent be expelled from the Party.

'Quite right too,' friends remarked, 'You did break the rules.'

The truth of the matter is that I have never felt represented or that I have any voice within our supposedly democratic system of governance. First past the post consolidates minority rule, allowing the single individual with the most votes in any constituency to win, regardless of however many votes add up against him or her. This has meant that for most of my life I have been governed by a party that has not had the support of the majority of the electorate. For example, Mrs Thatcher came to power in 1979 with 44 per cent of the vote and held on to power in the next three elections, even though on every occasion more than 50 per cent of the country voted for parties to the Left of her.

In 2019 the Tories won an eighty-seat majority with 43 per cent of the vote. They represent even less if you factor in the 30 per cent of eligible voters who did not bother to vote at all.[21] And given that by 2022 we were on our third prime minister in one year, the argument that it produces strong and stable government is laughable. Rather, it locks us into a two-party system. George Washington may have been a slave owner and led a settler colonialist government, but his farewell address to the US Congress was remarkably prescient when he warned that 'The alternate domination of one faction over another, sharpened by the spirit of revenge, natural to party dissension, which in different ages and countries has perpetrated the most horrid enormities, is itself a frightful despotism.' He went on to point out that it could result in a more formal and permanent despotism:

> The disorders and miseries, which result, gradually incline the minds of men to seek security and repose in the absolute power of an individual. [...] It agitates the Community with ill-founded jealousies and false alarms; kindles the animosity of one part against another, foments occasionally riot and insurrection. It opens the door to foreign influence and corruption.[22]

A two-party system in which most people feel their votes count for nothing does not produce a functioning democratic state. It produces Donald Trump, Jair Bolsonaro and Viktor Orbán. Proportional representation gives every vote equal value and demands compromise and consensus building that dilutes the power of any single party. But simply creating a more representative voting system is not enough.

* * *

When I lived at Greenham, I defined myself as an anarcho-pacifist. I share the primatologist Frans de Waal's view that human beings are basically cooperative, prosocial, empathic mammals who strive for fairness, want to avoid conflict, and desire to love and be loved.[23]

But my Gandhian ideal of living in a loose network of village cooperatives is not achievable in a world where inequality kills one person every four seconds and where the twenty richest billionaires in the world emit 8,000 times more carbon on average than the poorest billion.[24] Oxfam's devastating 2023 report, 'Survival of the Richest', showed that extreme poverty had increased for the first time in a quarter of a century. Of new wealth generated since 2020, 63 per cent has gone to the richest 1 per cent. It is not trickling down to the remaining 99 per cent. Trickle up would be a better term. For every dollar of new global wealth made by someone in the bottom 90 per cent, each billionaire gained US$1.7 million.[25] Nor can village cooperatives take on the divisive power of an unregulated social media which fosters division and hatred in pursuit of profit. In today's complex, globalised, increasingly urbanised and industrialised world, the only way to take on the plutocracies and technocracies that dominate our lives is through creating functioning democratic states that create just laws and adhere to them, both nationally and internationally.

And to be clear, a functioning democratic state is not one where the biggest concern is how the markets will respond to the undemocratic imposition of a new prime minister, rather than how people will. A functioning democratic state has not been bought by the fossil fuel industry, which then ensures that their lobbyists and proxies are appointed as that same unelected prime minister's special advisers.[26] Nor does it contain a legislative assembly whose members are there because of hereditary privilege, or cronyism. In a functioning democratic state, bankers are not more highly valued than nurses, and it would be unlikely that a prime minister, thrown out of his job as a liar, could earn more than £1 million for giving speeches in the following months,[27] while almost 4 million children in his country were living in poverty.[28]

In functioning democratic states, health, education and national infrastructure are public goods and services, not the means to private profit; and the most significant measures of 'development' are not based on material exchange, consumption and economic growth, but human wellbeing and the wellbeing of the natural world in which we are embedded, and upon which we depend.

I stress the word functioning. Functioning democracies confront and limit the power of those elites who wish to control them. This is a matter of life and death. As the Canadian social science professor, Kevin MacKay, explained in an essay written in 2018:

> Oligarchic control compromises a society's ability to make correct decisions in the face of existential threats. This explains a seeming paradox in which past civilizations have collapsed despite possessing the cultural and technological know-how needed to resolve their crises. The problem wasn't that they didn't understand the source of the threat or the way to avert it. The problem was that societal elites benefitted from the system's dysfunctions and prevented available solutions.[29]

That is exactly what is happening today. 'Fossil fuel producers and financiers have humanity by the throat', UN Secretary-General António Guterres warned in June 2022: 'For decades, many in the fossil fuel industry has invested heavily in pseudo-science and public relations—with a false narrative to minimize their responsibility for climate change and undermine ambitious climate policies.'[30]

Some may argue that corporations are now too powerful to take on, but as David Runciman points out in *How Democracy Ends*: 'No corporation, however rich and powerful can exist without the support of the state. Corporations are created in law, and they operate through the web of rules and regulations that the state provides to manage them.'[31]

If we are to confront the kidnapping of our democracies by corporate elites, a fairer voting system, combined with more devolved power, is essential but insufficient. I don't want to live in a society where people only possess power on election days, and, to quote another founding father of American democracy, 'After this, it is the property of their rulers, nor can they exercise or resume it, unless it is abused. It is of importance to circulate this idea, as it leads to order and good government.'[32]

On the contrary, it leads to chaos and corruption. If we want to live in a functioning democratic state, where the body that makes the regulations by which we should all abide does actually consult with us rather than battling for power every five years and then ignoring us, we have to go beyond representative democracy. We need a participatory system where politicians are not listening to paid lobbyists and a billionaire-controlled media, but give their own citizens an effective voice. This is why Extinction Rebellion's third demand: 'Decide together', and their call for a Citizens' Assembly, is to my mind the most important.

The idea is simple. A group of citizens are selected through a process of stratified random selection—called sortition—so that it is representative of the public at large in terms of class, ethnicity, gender and geography. This 'mini public' is then tasked with meeting regularly over a period of time to deliberate and discuss issues of public significance. They will listen to a wide range of expert opinion, examine written evidence and then submit their conclusions for action by government.

It's not a new idea. One of the pillars of Athenian democracy in the fifth century BCE was an Assembly open to all male citizens in which they could 'listen to, discuss, and vote on decrees that affected every aspect of Athenian life, both public and private, from financial matters to religious ones, from public festivals to war, from treaties with foreign powers to regulations governing ferry boats'.[33]

More recently, Porto Alegre in Brazil, a city of more than 1 million, established a tiered consultative process that allowed 17,000 people to participate in allocating the city budget. The results? A much more inclusive process, in which women, ethnic minorities and those with lower income were involved; less corruption and a more equitable distribution of resources in which health, education and the environment all benefited. Participatory budgeting in other communities and countries has had similar effects. As Rutger Bregman argues in *Humankind: A Hopeful History*, if people feel their contribution is valued and taken seriously, they want to participate; trust grows as does engagement and feelings of solidarity with the community as a whole.[34]

One of the best-known examples is the Irish Citizens' Assembly convened by the Irish government in 2016–17 to examine and make recommendations on the contentious issue of the Eighth Amendment to the Irish constitution, which had banned abortion since 1983. Ninety-nine randomly selected citizens met and deliberated for five sessions over six months. They listened to twenty-five experts and reviewed 300 submissions from members of the public and interest groups. They adhered to key principles to guide discussion. These included: 'openness of proceedings; fairness in how differing viewpoints were treated and of the quality of briefing material; equality of voice among members; efficiency; respect; and collegiality'. Many talked about how actively listening to expert opinion changed their minds.[35] In April 2017, 87 per cent of the Assembly members voted to amend the Eighth Amendment in some way, and two thirds of members voted for 'terminations without restrictions'. They then came up with recommendations that were put to the general public through a referendum. In May 2018, 66.4 per cent voted to repeal the ban on abortion, with the highest ever turnout for a vote on social issues.[36]

German-speaking Belgium has gone even further. They have established a permanent Citizens' Council and Assembly in which citizens chosen by lot, representative in terms of gender, age, education and place of residence, discuss and debate issues of importance chosen by the Council. Their recommendations are then made to the regional parliament, which has to respond fully to the recommendations made and discuss those areas where they disagree.[37]

Extinction Rebellion's 2019 action did lead to the government agreeing to commission a UK Climate Assembly. As a result, 108 randomly selected citizens, including the full range of opinion on climate issues, met six times in the spring of 2020 to deliberate and make recommendations in response to the question: 'How can the UK reduce greenhouse gas emissions to net zero by 2050?'

But unfortunately there were a number of limitations and failings in the process. The completely inadequate 2050 target date was not up for discussion, and the UK's international carbon emissions, such as those from the overseas transport and manufacture of goods for UK consumption, were not considered. The public were not engaged, and had no means by which to send comments. Most significantly, as the Assembly was commissioned by six select committees who play an advisory role, there was no requirement for the government to respond to or act on the recommendations.

Which is a pity, because the process itself was constructive, the participants were committed, enjoyed it and produced some good ideas. The 108 citizens were able to agree on key principles that underpinned all their recommendations. These included 'improved information and education for all on climate change; fairness, including across sectors, geographies, incomes and health; freedom and choice for individuals and local areas',[38] but not 'at the expense of taking the steps necessary to ensure a safe and healthy environment for future generations'; strong proactive leadership from government including the establishment of a cross-party consensus; and protection and restoration of the natural environment.[39] They then went on to make detailed recommendations in specific areas. Some are particularly eye-catching, such as 'follow the principle that the polluter should pay'.[40]

And when, at the end, individuals were invited to make additional recommendations outside the framing question, the Assembly passed some much more radical proposals, all of which had 89 per cent support or more. These included:

> Get to net zero without pushing our emissions to elsewhere in the world; more transparency in the relationship between big energy

companies and government; an independent neutral body that monitors and ensures progress to net zero, including citizens' assemblies and independent experts; move away from fossil fuels and transition to new energy sources.[41]

It would appear that when randomly selected citizens with widely diverging views are given the space to listen to a broad range of expert opinion; to think, discuss and deliberate without lobbying or interference, they can come up with good recommendations that make sense to the public as whole. In fact, according to an OECD report, public deliberative and participatory processes have been increasing at all levels of government across the world, particularly in the last decade. They cover a wide range of issues, with environment and health being among the most common. And in around two thirds of cases, at least half of the participants' recommendations are accepted by public authorities.[42]

It can be done. Extinction Rebellion now demands that the government must create and be led by the decisions of a Citizens' Assembly on climate and ecological justice, commissioned by parliament but run by an independent NGO. They argue that this is the only way to make the difficult long-term decisions needed to address the climate and ecological emergency in a fair way. I would go further. Why not demand a permanent assembly as in German-speaking Belgium? And why limit the topics? The climate and ecological crises are one terrifying aspect of the crisis in democracy as a whole. A Citizens' Assembly would be the ideal place to thoroughly discuss and make recommendations for constitutional reform, a just system of taxation, reparations to countries damaged by slavery in the past, and for the loss and damage caused by the climate and ecological crises today. What's not to like?

* * *

Something is shifting. In February 2022, Jan Power, a doctor colleague of mine active in Extinction Rebellion, was elected onto the local town council as a Green councillor. She won her place because the Liberal Democrats agreed to stand down and give her a clear run. She told me the story as we drank coffee near St John's Hall in Penzance, where the council held its meetings.

Jan has been an activist since the age of fourteen, when she campaigned against the inclusion of CFCs in aerosols in order to protect the ozone layer. After qualifying she joined an equal pay collective GP practice in Sheffield, one of three in the country at the time, where everyone from cleaner to doctor was paid equally and shared in decision-making. Along with others she took Sheffield Council to court to prevent them turning a precious area of South Yorkshire woodland into a clay pigeon shoot. They won. In 1996, she and her partner moved to Cornwall where she worked as a GP until her retirement. It was young 'Fridays for Future' activists who inspired her to join Extinction Rebellion in 2018. She had gone to support them on a strike outside County Hall:

> I was just humbled and blown away by these kids, how articulate they were and how passionate they were and how brave they were, because various county councillors were coming out and saying 'You should be in school' and being quite angry with them for what they were doing, and they were just explaining their purpose in a really powerful way and I stood there and thought I can't leave it to you to sort this mess out that's of our doing.

So she joined Extinction Rebellion and has been arrested some five times, being found guilty and fined on three occasions. But then she decided something else was needed. 'It's all very well going up to London and raising awareness. I think it's important. I also think it's very important to act in your community. This is my chosen home for the last twenty-six years. I started to feel increasingly strongly that I needed to be working from within, as well as without.'

And that is what she has done. In the year since being elected, Jan has introduced a 'Future Generations Pledge' that is made at the beginning of every meeting: 'It says that all our decisions have to take into account the wellbeing of future generations.' She saw a decision like this as having enormous importance: 'If you start thinking about what is in the interests of the generations not yet born, then it drives things like social justice and a Citizens' Assembly, so it's like the underpinning.' Meanwhile, she was ensuring that local environmental NGOs were involved in the council's climate emergency sub-committee. When we met, she was working on an ethical procurement policy which would affect pensions, insurance, investment and banking. She knew how small the town council was, but pointed to the ripple effect, not just locally but beyond. Other towns such as Brighton and Oxford were also considering the future generations pledge.[43]

Meanwhile, others are pushing local democracy in other ways. In November 2022 I joined a large gathering for a People's Assembly at County Hall in Truro, where Tories lead the council, having captured 54 per cent of the seats with 37.9 per cent of the vote. My friends David and Phil had already climbed on the roof and hung a banner from some scaffolding saying 'Emergency, Save Lives, Insulate Cornwall, Green Jobs, Not Rocket Science'.

Downstairs in the central lobby the rest of us, representing a range of different groups and issues, were there to discuss getting the council to actually act on the climate emergency it had declared in 2019, fuel poverty and the housing crisis. Cornwall is overwhelmed with second homes and Airbnb lets, but local people, nurses and teachers cannot find anywhere to rent.

The holding of open people's assemblies alongside full council meetings began in September 2021, when large numbers had occupied the space uninvited. The council's first response was to call the police. According to fellow activist Myghal Ryual, they allowed the meeting to continue for three hours. A dialogue was initiated with the council leader, who along with three of her colleagues met with four representatives from the Assembly.[44]

Now the Assemblies are held every couple of months alongside full council meetings. Some councillors pop in on their way to that meeting, and the results of the Assembly are passed on to the leadership. It is not a formal Citizens' Assembly and has no powers, but the enacting of open democracy in this manner possibly has an effect. In December 2022, the council approved the imposition of a 100 per cent council tax premium on second homes.[45]

Asmamaw shares an Amharic proverb: 'Drops of water break the rock.'

SHUT UP AND LISTEN, MARCH 1984[46]

There have been times last month in Newbury Magistrates' Court when I have felt like Alice at the trial of the Knave of Hearts. However, he at least had the benefit of trial by jury—albeit an odd one—and when the proceedings became too arbitrary and farcical for Alice to bear, she could leap to her feet, clap her hands and turn the whole charade into a pack of cards.

Greenham women are not so fortunate. Any who leap to their feet to complain of injustice may rapidly find themselves in the cells below: laughter is forbidden and I can only be grateful—given the frequency with which the Stipendiary Magistrate Mr Miller cleared the court—that he did not have the Queen of Hearts' power to behead us all as well.

A neatly dressed, clean-shaven, rather dapper man, he arrived at the beginning of January and, to the delight of *The Daily Telegraph*, began to despatch high-speed justice in order to clear the back-log of cases hanging over from the actions at Greenham during the preceding few months. Not only did the rate of cases increase (on occasion 78 were heard in a single day), but so did the rate of acquittals. In his month of residence Miller acquitted some 20 women on the basis of inadequate police evidence.

There was a catch, however. Nearly every one of these acquittals was given to women represented by counsel. Yet legal aid has become increasingly difficult to obtain. ('There is nothing complicated about the British legal system', Miller told an American woman in turning down her application.) Help in the form of a 'McKenzie friend' (a lay adviser who can sit with the defendant) has also been prohibited.

To the unknowing defendants, the procedure appears to offer a most summary sort of justice. 'Shut up and listen!' Miller told a 20-year-old defendant appearing in a Magistrate's court for the first time when she asked for such help. He repeated the admonition twice during the trial when she again asked, and then, impatient at the way in which she cross-examined the police witnesses, dismissed them without asking if she had finished. In mid-trial her co-defendant, also 20 and on her first appearance, requested an adjournment to seek representation 'because I feel totally inadequate'. This application was again turned down and the court cleared of women murmuring support from the back. On this occasion both women were found guilty of criminal damage and, in spite of asking for time to pay their £35 fines,[47] were jailed for 14 days for failing to provide an adequate address.

At least six women have so far been jailed for refusing to give any address other than Greenham Common Women's Peace Camp when asking for time to pay their fines. Miller repeatedly stated that he wouldn't accept it as a place of abode.

'Why is she so insistent that Greenham Common is her address?' he asked Katherine Cronin, defending Vicki McCafferty. 'She must have relations, a place where she keeps her personal belongings? What is so special about it?'

'She is not required by statute to give her parents' address,' Ms Cronin replied, 'not even her only address, but simply one at which you can be assured that she can be found.'

'But it might disappear overnight!' responded Miller. Ms Cronin went on to point out that the same might apply to squats, caravans and bedsits, all of which have been accepted by the courts and that he 'did not have the right' to turn down an address that both the DHSS and the Post Office found acceptable. Moreover, on the basis of mail received at the camp and her presence on the Electoral Register, Judge Blomfield had ruled on appeal at the Reading Crown Court on 13 January that Greenham Common could, in the case of a particular woman, be regarded as an abode.

This was not enough for Mr Miller. 'I have had a personal telephone conversation with the Judge,' he told us, 'and he informed me that his judgment was totally misreported.' Vicki McCafferty was given 14 days.

For many women, who have given up rented accommodation in order to live permanently at the camp and who have no contact with their families, this judgment presents an impossible dilemma. Either they must give a false address which is acceptable to the Magistrate, or they have to leave the camp and live somewhere else. The result is a rather subtle form of eviction. While many small-scale evictions continue every day around the base, this is one form that we are not prepared to accept.

The sheer number of cases going before the Stipendiary Magistrates has created a new set of problems. The range of charges is small and in prosecution after prosecution the police evidence is similar. How, then, can a single Magistrate hearing all these cases possibly bring to each the fresh and unbiased mind that justice requires? What has occurred is that decisions made on the law—and probably the facts also—are merely reiterated time after time. In dismissing Owen Green, an expert witness for the defence on cruise missiles, Miller said, 'We all live in the same world and I am perfectly aware of these facts.'

The anxieties about bias extend also to the lay benches. Not only have they listened to over 1,000 cases from Greenham Common, but they are also local ratepayers and some of them serve on the Police Authority. Policing costs stand at well over

£3 million and there is extreme hostility to the camp in Newbury, fed by the press and anxieties about an increase in the rates. Given these circumstances it is hard to believe, as one woman Magistrate claimed, that all cases are heard 'without fear or favour, malice or ill-will'. Certainly, their acquittal rate—under 1% at the time of writing—does not seem to bear this out. Despite this, all applications for these benches to dismiss themselves, or for cases to be transferred to another jurisdiction, have been refused.

In the face of these difficulties women have become more determined to have their cases tried by juries. Certain charges, such as criminal damage to property in excess of £200, or 'going equipped to cause criminal damage', give this right. One group of women involved in taking down the fence on 13 July last year, is to appear in Reading Crown Court.

Since then, well over 200 women have been charged with offences which give them a right to elect for jury trial. Yet in all but three cases these charges have been dropped or modified with the effect of removing that right. Seven women charged jointly with causing over £10,000 worth of damage by painting the American spy-plane *Blackbird* last summer found that their charges were abandoned, without explanation, on their first appearance at court. When eleven women were charged jointly with 'going equipped' to paint the runway on 30 October, they were told, only four days before the committal, that the charges had been changed to 'causing £25 worth of damage' and that they would be tried summarily. Almost all the women arrested on 29 October were charged with 'going equipped' and again in almost all cases this charge has been unaccountably dropped.

Mr Barr, the new Stipendiary, has told women that the court is not a political forum. However, having watched the contortions of the prosecution in their charging strategies over the past few months, it is hard to escape the conclusion that we are being subjected to political manipulation designed to contain and control. It would not be for the first time. After the spectacular acquittals in the Bristol Riot cases, there seems to have been a marked tendency towards summary charging in 'public order' cases, so defendants cannot bring out what they believe to be the politics of the case.

The suffragettes had a similar experience. On Black Friday, 18 November 1910, 115 women and four men were arrested when a deputation of women marched on Parliament and for six hours were repeatedly, and with great violence, thrown back by the police. The following day those arrested found that, instead of being able to plead their cases, they were merely brought before the bench and reprimanded. Winston Churchill, the Home Secretary, had decided that 'no public advantage would be gained by proceeding with the prosecution'.

Justice, as I understand it, is not about 'public advantage' (whatever that is), nor is it about speed, efficiency and the saving of costs. It is about individual rights and fair and equal treatment before the law, whatever one's financial state or political beliefs. It seems ironic that the arrival of cruise missiles—deployed, we are told, in order to protect our liberties—should trigger this particular round of arbitrary and discriminatory treatment in the courtroom. Yet it is quite apparent that in spite of the increasingly heavy sentences, ranging from seven days to five months, and the increasing numbers of women in prison (30 at the time of writing), the courts are failing totally in one important function, which is the prevention of crime as they define it.

'I did cut down huge stretches of fence,' Katrina said in her defence, 'it was one of the most powerful experiences of my life and I will do it again. So, I can see that you are in a difficult position because, of course, whatever you do to me I shall continue to keep on doing non-violent action for as long as I believe it is necessary. Your problem is how to stop me.'

Katrina paid her fine. 'Some of us,' she remarked, needed to stay 'free and fighting'.

5

LAW OR JUSTICE?

Diary: 14 November 2022, City of London Magistrates' Court

'I don't want to be arrested; I accept it as a consequence. I am heavily invested in the rule of law and I do this to protect the law.'

Bristol GP Dr Patrick Hart is defending himself against the charge of breaching the Section 14 direction of the 1986 Public Order Act, that was imposed when some thirty of us had sat down on Lambeth Bridge the previous April. He is one of the seven who refused to move when asked. Eight months later I am back in a windowless room on the second floor of the City of London Magistrates' Court. The seven defendants sit to one side. Supporters and visitors sit on hard chairs behind the prosecution and barrister, opposite Judge Robinson.

It has been a long day. The prosecution put their case this morning that the Section 14 was properly applied, and was a necessary and proportionate response to prevent serious disruption to the community. This was largely through the testimony of Superintendent Jason Stuart, a Bronze commander who had three years' experience of public order policing, including football matches, festivals and all of the Extinction Rebellion campaigns.

Superintendent Stuart explained in detail what he had done that afternoon after arriving at Lambeth Bridge around 2 p.m. Initially he had allowed all the protesters to sit because he wanted to strike the right balance between their right to protest, the rights of others to go about their lives, and the risk of serious disruption. That changed when he heard from a Silver commander that there were serious impacts on traffic in other parts of London: 'Some ninety-two buses had been affected.' He then explained that he sent some ten to twenty officers out among the demonstrators to ask them their plans and whether they would move.

'We walked through the crowds to understand their intentions, to see if they would move to the pavement, they said hello, but would not engage on their plans.'

This was odd, because I had been on Lambeth Bridge all afternoon myself and did not witness any police officers chatting to any of us, prior to the actual imposition of the Section 14, which happened very quickly after the health workers sat down. This was in marked contrast to the previous day when large numbers of Extinction Rebellion protesters had arrived in Trafalgar Square in the early afternoon and small groups had blocked all the surrounding roads. The police had redirected the traffic, moved around among the demonstrators, and the atmosphere was so peaceful that I and some friends had played a game of cricket across one junction. When I left the square at around 7 in the evening, the police were just beginning to talk to protesters about imposing a Section 14.

Everyone in the Dock challenged him directly when given the chance to cross-examine: 'There was no engagement prior to your declaration of a Section 14,' consultant obstetrician Alice Clack said. 'I have no recollection of this discussion.'

But Superintendent Stuart was clear: 'We had to be sure we explored the least intrusive options. We really would have facilitated protest anywhere. There was one-to-one engagement, but my officers told me they wanted to be arrested. They weren't going to move.'

Patrick's focus this afternoon is on whether the conditions imposed on their disruptive protest were proportionate, given the level of disruption likely to be faced in the future.

'If it's the death of millions, the loss of everything we care about, in particular the loss of law and order?' He had explained how working as a doctor in East Africa and witnessing children dying of malnutrition had made him realise that he had to do more.

'I did every legally acceptable thing I could think of to effect change. [...] I absolutely accept I was causing disruption.'

'There is no evidence that this sort of protest is more effective than rallies. These protests have taken place since 2019 with no effect', says the prosecutor.

'Actually, you are wrong. In 2019 parliament declared a Climate Emergency and then we had the pandemic,' Patrick responds. 'Look at the suffragettes, anti-Apartheid, civil rights, name me one that has not used disruption.'

'Gandhi was not living in a democratic society', the prosecutor counters.

'This is an unprecedented situation. Within our democratic society our politics have failed. The UN Secretary-General, António Guterres, said last week: "The climate crisis is the fight of our lives— and we are losing. But one thing is certain: those that give up are sure to lose." This is our best attempt at salvaging something. If you are right, the result will be hell.'

15 November 2022, City of London Magistrates' Court

My other six colleagues presented their defence today. One was represented, the other five defended themselves. As I listened, I was jealous once again. Yesterday, the prosecutor, on opening the case, argued that 'the defence of necessity is not appropriate in cases of protest. There is no connection between what the protesters are doing and the harm they say they are trying to prevent.' But in marked contrast to

Stipendiary Barr all those years ago, Judge Robinson gave each defendant all the time they needed to explain both their personal motivations and their memories of what had happened on the day.

Until this hearing I had not realised how many of my colleagues, like Patrick, had worked in the Global South. GP David McKelvey grew up in a mixed-race neighbourhood in South Africa and worked in Tanzania in the 1990s. It was pictures of the loss of snow on Kilimanjaro, and the realisation of what that meant in terms of water shortages for those living on the plains below, that made him join Doctors for Extinction Rebellion. Alice spent three years in Sierra Leone, Liberia and Gambia and saw climate change undermining every gain achieved. She worked in Yemen for three months last summer and witnessed the effects of 40–50°C temperatures and flash flooding on those already suffering from war. She began to weep as she gave her testimony and I watched Judge Robinson tell her gently to sit and take the time she needed.

'Our planet is in a premorbid state; as a doctor I have a duty to tell the truth and protect patients', she continued. In different ways all made the point that the climate crisis was a health crisis, and they regarded action to address it as part of their professional duties. 'The GMC tells us it is our duty to promote and protect the health of the public', said David. Specialist nurse Anna Bunten explained: 'My professional body is clear: if we see a threat to patient safety or public health we must respond', before going on to talk about her brother-in-law who had died of bronchiectasis caused by air pollution, which according to WHO is making air unsafe to breathe in 90 per cent of the world's cities.

Chris Newman said he became involved in activism 'when I realised the risks to society with climate change were far greater than the risks to the patients that I was referring for urgent care'. He went on to explain:

> *As a GP I think a lot about risk ... there is a disparity between the individual and collective risk we are prepared to tolerate. If the risk to the*

individual from their illness is higher than 3 per cent, we must refer urgently. For example, a three-year-old girl with persistently enlarged lymph glands must be seen within two weeks. Yet an emissions gap report by the UN, published in 2021, said there is a 50 per cent risk of the average global temperature exceeding 1.5°C by 2041.

He quoted a joint press release by The New England Journal, The Lancet *and* The British Medical Journal, *three of the top four medical journals in the world, which along with 200 other journals stated: 'The science is unequivocal: a global increase of 1.5°C above the pre-industrial average and the continued loss of biodiversity risk catastrophic harm to health that will be impossible to reverse.'*[1]

All of them had tried conventional legal actions and felt they were not enough. Chris quoted a Lancet *editorial on the role of provocative protest that had been published the previous week:*

There is some evidence that disruptive or radical nonviolent actions, within the context of a wider movement that uses a mix of protest and lobbying approaches, is successful at garnering public attention for a cause, with no loss of, or possibly even increased, support for a social movement's overall goals.[2]

The defendants also all agreed that there had been no interaction between the police and demonstrators earlier in the afternoon. None of them had told any officer that they 'wanted to be arrested'; and no one had discussed alternative options for continuing the protest in a less disruptive way. 'If we had discussed it together, we might have been prepared to let some buses through,' Chris said, 'the idea of letting buses through is a good one.'

The imposition of a Section 14 was simply announced after the other demonstrators had left. 'Some twenty-two minutes later, officers approached us to give us the option to leave or be arrested.'

'We were trying to navigate a space, we wanted to be disruptive enough to raise our voices and be heard,' David explained. 'There is a difference between wanting to be arrested and being prepared to be

arrested.' He pointed out that they had not come equipped with glue or warm clothes. Moreover, as their action outside the Treasury three days earlier had shown, given reasonable time, they could disband and move on. 'In that twenty-two-minute window there was no discussion with us to allow us to discuss other options; if they had allowed us a bit longer, I honestly think it might have ended differently.'

The prosecutor didn't agree: 'The prosecution case is that liaison did engage before the Section 14 was applied, and everyone saying it didn't happen is lying, or was so disengaged they did not notice.'

'I am not going to dispute the science, no reasonable person would do that', he assured us, but once again he seemed anxious to establish that this kind of protest was ineffective. 'If this extraordinary summer with high temperatures and wildfires didn't convince people, there is no way your protest could affect people', he told Anna. He made a similar point to Alice: 'It must have been obvious that this government isn't amenable to this kind of protest.'

I am curious as to why he wanted to push this point. Is it perhaps because, if he does agree with the science, and recognises the threat, he then needs to rationalise why he is prosecuting health workers trying to do something about it, and not joining in the protests himself?

Retired GP Mark Russell challenged him, describing his own involvement in an Extinction Rebellion action outside the BBC in October 2019, protesting their failure to adequately cover the climate crisis: 'We said we were not going anywhere and they invited us in. That same month there was a dramatic change in the conversation and their coverage improved.'

But this had no impact on the prosecutor's closing remarks, where he reiterated again that he had 'no doubt in the sincerity of their beliefs or the justice of their cause. No reasonable person would dispute the science, but their actions have no chance of shifting the government stance or of changing anyone's mind.' Moreover, the defendants could not argue necessity, because they had 'to be acting to prevent immediate loss of life or injury' and no one 'had been able to draw the nexus between their actions and the dangers they hope to avert'.

Justice Roberts came back after some ninety minutes of consideration. In a quiet, dispassionate voice he gave a detailed summary of the key points of the case. He agreed with the prosecution that the defendants had failed to show evidence of the necessity of their action. However, he had been impressed 'with the integrity and rationality of their beliefs', and he had found their evidence 'highly moving'. Moreover, 'no one had recall of officers coming to discuss alternatives'. Thus, he could not 'find any evidence beyond reasonable doubt that Superintendent Stuart engaged in a meaningful way to persuade them of alternatives. [...] I cannot be satisfied in this case that a proper assessment necessary for the imposition of Section 14 was made. I find the case not proven.' We rose to our feet as he left the courtroom. Chris looked stunned and wept as he leant against David's shoulder. Mark was crying, others were cheering and hugging. All were free to go.

* * *

Can law-breaking by climate activists be used to further justice, as John Rawls suggests? It would appear so. As noted above, this was not the first acquittal of climate activists occurring at this time. In October 2022, five scientists had been found not guilty of causing criminal damage at the Department for Business, Energy and Industrial Strategy (BEIS) building in April 2021. They were part of a group of nine who had pasted scientific papers and sprayed soluble chalk signs onto the windows, as well as glueing themselves to the building. And in this case the acquittal was not on technical grounds, but because the magistrates agreed with the argument put forward by the defence, that conviction would not be a proportionate response to the action, as the damage caused was minor and temporary, and the defendants were exercising their right to protest under the Human Rights Act 1998. There was no case to answer.

However, the arbitrariness of rulings in Magistrates' Courts was exemplified by the fact that a month earlier a different district judge had found the other four scientists involved in the same action at BEIS guilty of criminal damage, having refused to allow arguments based either on the necessity to prevent greater harm, or on the defendants' right to freedom of expression under the Human Rights Act.[3]

Acquittals were also occurring in the Crown Court, not just for criminal damage (as mentioned in Chapter 3). Six activists who had climbed onto the Docklands Light Railway carriage roofs at Canary Wharf in April 2019 and hung banners saying 'Business As Usual=Death' and 'Don't Jail the Canaries', were acquitted of obstruction in December 2021. The jury agreed that they had lawful excuse because of government inaction on the climate crisis.[4] A few weeks later, in January 2022, three activists who had been involved in obstructing the DLR in Shadwell in October 2019 were also acquitted. In his summing up, Judge Silas Reid instructed the jury to decide whether the actions of the defendants and the disruption they caused was 'proportionate' with their right to protest. Motivation and beliefs were irrelevant:

> Intentionally disruptive action, even that which has more than a minimal impact on the rights of other people, need not result in a conviction. It is all a matter of fact and degree. It requires balancing the defendants' rights to free speech and to protest, as against the rights and freedoms of other people from being obstructed from going about their daily business and the importance and need in society to prevent disorder and secure public safety.[5]

The jury came down on the side of protest. Judge Reid was clearly taking into account the Ziegler ruling, in which the Supreme Court, in June 2021, had made clear that the right to protest, as enshrined in Articles 10 and 11 (the rights to freedom of expression and to freedom of assembly) of the European Convention on Human Rights, 'extends to a protest which takes the form of intentional disruption obstructing others. However, the extent of the disruption and whether it is intentional are relevant factors in the assessment of proportionality.' This meant 'considering whether the public authority's interference with the protestors' rights was proportionate. If it was not proportionate, such that the interference was unlawful, the protestor will have a statutory defence of lawful excuse.'[6]

Ziegler is not just a ruling. Nora Ziegler is a person, one of four Christian peace activists who on 'No Faith in War' day on 5 September 2017, lay down and locked on to boxes to obstruct one lane of the entrance road to the ExCeL centre in East London, where preparations were being made to host the Defence and Security Equipment International (DSEI) exhibition, otherwise known as an arms fair. In an interview about the action, she described herself as a Christian and anarchist who had lived and worked with the Catholic Worker Movement in London for some years, supporting refugees and asylum seekers and the homeless. She saw the day as an opportunity to 'come together and express together that we believe in God, we believe in love, we believe in peace, and we think that's incompatible with the arms trade, we think that's incompatible with borders, with violence, with war'. Her hope was that a number of different actions would take place: 'The idea of the action, of the lock-ons, was that we would hold a space so that people wouldn't be cleared away, so that the police wouldn't be able to disrupt the prayerful events that were happening.' The police arrested them within eleven minutes of initiating the action.[7]

They were charged with wilful obstruction of the highway without lawful authority or excuse under Section 137(1) of the 1980 Highways Act, but in February 2018 the District Magistrate found them not guilty, because 'the district judge had regard to the appellants' article 10 and 11 ECHR rights, and concluded that the prosecution had not proved that the appellants' obstruction of the highway, which it found to be limited, targeted, and peaceful, was unreasonable'.[8] However, the prosecution appealed, and in 2019 the High Court concluded that the district judge had failed to strike a fair balance between the interests of the protesters and those of other members of the public, because 'the ability of other members of the public to go about their lawful business was completely prevented by the physical conduct of these Respondents for a significant period of time'. They overturned those acquittals. So, the four appealed to the Supreme Court, and on 25 June 2021, that court ordered that their convictions be set aside. Raj Chada, the solicitor who had acted for the four initially and advised them throughout, summed up in an interview with *Peace News* what this judgment meant: 'the state, and that included the court, had to ensure that a conviction was a proportionate interference with the defendant's right to free speech. [...] And, if they weren't sure that [conviction] was a proportionate interference, they should acquit.'[9]

The issue of proportionality had never been expressed this clearly before. The judgment also clarified that proportionality and the degree to which the state authorities were restricting the protesters' rights had to be considered at each stage of a protest: during arrest, conviction, prosecution and sentencing. This began with the arresting officers considering whether they had given the protest enough time. Had they offered or considered alternatives? This was something Superintendent Stuart had been so keen to establish had been done on Lambeth Bridge in the doctors' case, but which was disputed by my friends.

The Ziegler ruling did have a marked effect on the policing of demonstrations, as I witnessed outside the Treasury, and again in Trafalgar Square in April 2022. It also had an immediate impact on what happened in court. In July 2021, Ben Benatt, an environmental scientist from Brighton, used this defence when on trial in the City of London Magistrates' Court for sitting in the road the previous September. District Judge Cohen agreed, explaining that 'the limited nature of the obstruction (both temporally and geographically), its peaceful nature, the importance of its objective and the genuineness of Benatt's beliefs' meant that he was not satisfied conviction would be 'necessary, and proportionate'.[10]

In January 2022, one of the arguments put by the defence in the Colston Four case was that because of this ruling, conviction would be a disproportionate infringement of the defendants' rights under Articles 9, 10 and 11 of the Human Rights Act 1998. This had possibly contributed to the jury deciding to acquit (see Chapter 3).[11] The following March, the High Court ruled that the Metropolitan Police had acted unlawfully when they prevented the organisation Reclaim These Streets from planning a vigil for Sarah Everard, the young woman kidnapped, raped and murdered by serving Metropolitan Police officer, Wayne Couzens, a year earlier. The judges, citing Ziegler, agreed with the organisers that the Met, in vetoing the vigil because of Covid restrictions, had not considered the organisers' rights under Articles 10 and 11, meaning they had lawful excuse.[12]

But the Attorney General Suella Braverman was not happy.[13] In April 2022 she made the unusual decision to refer the acquittal of the Colston Four to the Court of Appeal. This was not to reverse that decision, but to obtain legal clarification on two points. First, can those who commit criminal damage use the human rights defence? Second, do juries have the power to decide if a criminal conviction is a proportionate interference with a protester's human rights? Apparently, this questioning of the ability of juries to make complex value judgments, as they have been doing for some hundreds of years, was not in any way politically motivated by her concern about outcomes, but simply a wish 'to clarify the points of law raised in these cases for the future'.

The clarification was provided by the Court of Appeal in September 2022. Protesters could not use a human rights defence if their protest was considered violent, or not peaceful, or if they had caused significant damage to property: 'prosecution and conviction for causing significant damage to property, even if inflicted in a way which is "peaceful", could not be disproportionate in Convention terms'. So, although a judge may not direct a jury to convict, he could withdraw the issue of proportionality from the jury's consideration in these circumstances. However, in the 'limited circumstances' of very minor or temporary damage, 'a conviction may not be a proportionate response in the context of protest'.[14]

I imagine that Braverman was not best pleased when the magistrate took the judges' advice a few weeks later and acquitted those five scientists for minor and temporary damage to the BEIS offices. Nor can she have been happy when on the same day, nine other Extinction Rebellion activists who had been arrested on charges of aggravated trespass after entering Shell headquarters and glueing themselves to the furniture, also had their charges dropped.[15]

The acquittals have continued. In October 2022, six Insulate Britain activists were acquitted of multiple charges of wilful obstruction of the highway the previous year, after a magistrate in Horsham told them that sitting on the M25 for a short period was morally right and proportionate.[16] In December 2022, a jury took forty minutes to unanimously decide that three Insulate Britain activists who had blocked Bishopsgate in October 2021 were not guilty of causing a public nuisance. This was in spite of Judge Silas Reid, on this occasion, stating that they had no defence in law, and barring them from referring to the climate crisis, insulation or fuel poverty during their defence. Judge Reid had no more success when he issued the same instructions to four Insulate Britain defendants in January 2023. This time the jury took four hours to decide to unanimously acquit all of them of causing a public nuisance.[17]

No doubt as winter took hold and energy prices rose, some juries were well able to work out for themselves who was causing a public nuisance and where justice lay. As Gwen Harrison, one of the defendants in the December 2022 trial put it: 'By failing to insulate Britain's cold and leaky homes and allowing greedy energy companies to profiteer at our expense this government is sentencing millions of families to a winter of misery and starvation, in which thousands will die of the cold.'[18]

Appeal Court judges also appear to regard the climate emergency as relevant. In February 2023, the four scientists whom a magistrate had found guilty of criminal damage at BEIS, appealed at Southwark Crown Court and had their convictions overturned. Judge Rimmer pointed out that 'the protest was directly focused on the responsibilities of His Majesty's government for the climate emergency and in particular its role in issuing new oil and gas licences', and that the prosecution had produced 'scant' evidence of criminal damage. Judge Rimmer went on to make clear references to belief and motivation in his conclusion, stating that 'the protestors held heartfelt and real concerns about climate change and these are very important issues'.[19]

But there have been many convictions. Mark Ovland and James Mee, the two activists who climbed on top of that DLR train in Canning Town in October 2019, sparking a mini-riot, were found guilty and sentenced to a twelve-month community order in March 2022. Justice Silas Reid—yes, the same—had once again blocked the use of the proportionality argument, and told the jury they had no defence in law. But the jury foreman was given permission to read out a statement, which said: 'We, your Jury, want to take this opportunity to let the Court and the defendants know that we all wholeheartedly support the cause. Our verdict is based purely on the law rather than personal feelings about climate change.'[20]

By February 2023, at least 2,000 activists, mostly from Insulate Britain and Just Stop Oil, had been arrested for protest-related offences, and some 138 had spent time in jail, many without trial. This included activists like Roger Hallam, who spent four months in prison on remand, and Louis McKechnie, who was remanded in custody for three months before being sentenced to six weeks for tying himself to that goal post. He was then jailed for a further three weeks for glueing himself to the frame of an oil painting at the Courtauld Gallery. At the time of writing, Phoebe Plummer and Anna Holland are on bail until 2024 for throwing that soup at the Van Gogh painting. They are banned from entering any galleries or carrying glue in a public place.[21]

But even when climate activists have been found guilty of breaking the law, judges have spoken out. In May 2021, District Judge Bisgrove found former Greenham woman and Ploughshares activist, Rowan Tilly, guilty of blocking the highway in October 2019, because she had knelt down at the end of Regent Street with a placard saying 'Love and Grief for the Earth' and refused to move when asked. However, he drew attention to the changes achieved by previous unlawful actions, notably the US and South African civil rights movements, the suffragettes and the anti-nuclear testing protests, and praised Rowan for being honest and sincere. He said her deeds were 'noble' and gave her an absolute discharge.[22] A year later, District Judge Stephen Leake fined twelve Insulate Britain activists for blocking some 18,000 vehicles on the M25 because they had caused significant disruption, and he had sworn a judicial oath to 'apply the law'. But at the same time, he told them that they had inspired him and that 'personally I intend to do what I can to reduce my own impact on the planet, so to that extent your voices are certainly heard'.[23]

When Insulate Britain activist Christian Rowe was acquitted in January 2023, he warned against complacency: 'It just means that the police and the prosecutors will adapt their tactics and the government their laws for ensuring the punishment of future protesters, whom this government is intent on silencing.'[24]

Certainly, if governments feel that judges and juries are not putting enough nonviolent protesters in jail, and that the deterrent effect of current legislation is not working, there is another approach: introduce new legislation to criminalise the protesters.

It has been done repeatedly, as I had detailed in my lectures to those officers at the Royal College of Defence Studies in 1989. It was happening at that time. After more than 100 charges of riot and more than 500 charges of unlawful assembly were dismissed against striking miners, the 1986 Public Order Act was passed into law. This abolished those old common law offences and brought in new restrictions, including the Section 14 direction and the criminalisation of any behaviour that could be construed as harassment, alarm or distress by someone observing it. Some of the first to be arrested under the Public Order Act were two young people who put up an insulting poster of Mrs Thatcher, although the only person in the vicinity who could be 'insulted' was the arresting officer who was laughing at the time. Their charges were dismissed. The dangerous young man who made a V-sign and said a rude word outside No. 10 Downing Street was not so lucky.[25] I asked then what sort of government needed to criminalise words and gestures and how was it possible to have a demonstration that no one found disturbing in any way?

I was asking similar questions in an open letter at the end of 2021. A year earlier, the then Home Secretary Priti Patel, in response to the blossoming of nonviolent protests such as those by Black Lives Matter and Extinction Rebellion, had introduced the Police, Crime, Sentencing and Courts Bill, which would update the 1986 Public Order Act. Among the many new powers suggested were restrictions on the rights of travellers to camp, new powers to restrict and shut down peaceful assemblies if they were deemed to be causing 'serious disruption', and the right to stop and search without any cause.

The Bill appeared to have been inspired by a 2019 report entitled 'Extremism Rebellion' published by the conservative think tank Policy Exchange, which called for the law to be 'urgently reformed in order to strengthen the ability of police to place restrictions on planned protest and deal more effectively with mass law-breaking tactics'.[26] 'It appears that the Policing Bill is stained with the grubby, oil-soaked hands of the fossil fuel lobby. And no wonder—this cracks down on the fundamental rights of protestors to challenge the very climate-wrecking policies espoused by this downright dangerous industry', commented Green Party MP Caroline Lucas, when Open Democracy revealed that the American arm of this think tank had received a $30,000 donation from ExxonMobil corporation in 2017, because they 'assess public policy alternatives on issues of importance to the petroleum and petrochemical industries'.[27]

In the same report, Policy Exchange had also labelled Extinction Rebellion an extremist organisation. A few months later, counter-terrorism police in the South East briefly placed Extinction Rebellion on its list of extremist ideologies in an official guide entitled 'Safeguarding Young People and Adults from Ideological Extremism', produced for teachers and government organisations. Apparently, simply hearing young people speak in 'strong or emotive terms about environmental issues like climate change, ecology, species extinction, fracking, airport expansion or pollution' was a sign of potential radicalisation and warranted referral to the government's 'Prevent' programme.[28] As a child psychiatrist, I took part in regular Prevent training, and had always felt uncomfortable with it. When I read the guidance, I did briefly wonder if I should turn myself in. The designation was dropped after the government's Commission for Countering Extremism insisted that Extinction Rebellion should not be considered an extremist group. But in September 2020, shortly after an action where Extinction Rebellion had temporarily blockaded three Murdoch-owned Newscorp printworks in Hertfordshire, Merseyside and North Lanarkshire, delaying newspaper distribution for one day,[29] Priti Patel made her own views about Extinction Rebellion clear in a speech to the Police Superintendents' Association conference. Describing XR protesters as 'eco-crusaders turned criminals', she denounced the action against Murdoch's printworks as 'a shameful attack on our way of life, our economy and the livelihoods of the hard-working majority'. She continued:

> I refuse point-blank to allow that kind of anarchy on our streets. And I am right behind you as you bring the full might of the law down upon that selfish minority behind it. The very criminals who disrupt our free society must be stopped. Together we must all stand firm against the guerrilla tactics of Extinction Rebellion.[30]

Our open letter from mental health professionals was a way of responding. It was published online just before the House of Lords had their third reading of the Police, Crime, Sentencing and Courts Bill. It focused on the damage the Bill would do both to democracy and the mental health of young people—which the government said was a priority. We pointed to government guidance which itself stated that involving young people in decision-making could 'benefit their mental health and wellbeing', and we noted that 'Engaging in nonviolent protest is a democratic right that is part of such involvement, and restricting it in the manner envisaged in this Bill will further erode young people's trust in politicians, and their belief that their voices are heard, respected and matter.'

We cited the research on 10,000 16-to-25-year-olds from both the Global South and North, which found that almost six out of ten young people were very or extremely worried about climate change, and that these anxieties correlated with the perceived inadequate government response and associated feelings of betrayal.[31] Research had also shown that youth engagement in activism was one antidote to despair, and increased children's positive skills and capacities.[32] We quoted the COP26 president, Alok Sharma, stating that 'the actions and scrutiny of young people are key to us keeping 1.5 alive and creating a net-zero future', before concluding:

> Yet it is at this terrifying moment that the Government chooses to introduce new draconian legislation that will criminalise young activists engaging in nonviolent protest. [...] One result of this legislation, if passed, will be that young people face a choice between being intimidated into inaction and isolation, or possibly criminalised if they choose to act. Some may deliberately choose to escalate their actions to be more disruptive and possibly violent, given the severe consequences for even minor nonviolent activity. Whichever choice they make it will have profound consequences for their mental wellbeing and the wellbeing of the wider community.

More than 350 academics and clinicians signed. There was an article in *The Guardian* and interviews on the radio.[33] It was just one of multiple open letters and protests made at the time by police chiefs, lawyers and academics, among others. Just prior to the Lords' third debate in mid-January, I joined a digital action. I was amazed to get through directly to two peers and to be able to discuss the letter. One of the organisers wrote to me later saying that I was just one of some 2,000 who phoned, tweeted, or wrote and sent some 13,500 emails, and that peers had told them they had 'never seen such an outpouring around one bill before and thanking them for taking a part in the democratic process. [...] Despite how flawed and disheartening the political system can be, the fact that so many people overcame the challenge of even confronting it, and shared and took part in this action is incredibly inspiring.'

I was in Ethiopia on the day of the actual debate, and followed it sitting in my small bed and breakfast in Addis, looking at a twitter thread from Jessica Metheringham @QuakerDissent. She was sitting in the gallery of the House of Lords providing a blow-by-blow account of proceedings.[34] The irony of relying on unelected peers to challenge this legislation, rather than my own

MP who completely supported it, was not lost on me. Jessica tweeted a picture of Friends engaged in a noisy protest outside parliament: one held a sign saying 'Quakers quiet when we meet. Noisy on the street.' Inside, Lord Coaker (Lab) asked: 'Has democracy collapsed in the face of noisy protests? No, and never has any government sought to restrict it.' Conservative peer Lord Cormack supported the amendment to remove the proposed ban on noisy protests, pointing out that 'a Bill that is injected with a dose of stupidity is not a very good Bill'. Fellow Conservative peer Lord Deben agreed: 'Dissent and protest are essential parts of democracy. The government has produced no good reason for this legislation. Some protests are embarrassing, but these amendments go far too far. I need the opportunity to dissent and protest.'

As for restricting protests outside parliament, Viscount Colville pointed out:

> This is the mother of parliaments, outside which voters should gather to speak truth to power and where we, the parliamentarians who make the law, should hear them loud and clear. At this time, when politicians are seen to be out of touch with the feelings of the people, it is unconscionable that the House should pass a law shielding us from hearing what they have to say.[35]

And the Bishop of Leeds condemned the whole Bill: 'If this Bill goes through, we'll need to remove the statues of Gandhi and Nelson Mandela across the road. We can't laud them later when we would have condemned them.'[36]

The Lords were able to throw out Patel's last-minute amendments that criminalised locking-on and interference with public infrastructure, as well as those introducing new stop-and-search powers and the serious disruption prevention orders. But in late April 2022, when the Bill passed into law, it still gave the police new powers to stop noisy protests, restrict public assemblies, arrest those engaged in one-person protests, as well as creating a new offence of wilful obstruction of the highway. Soon after,

Asmamaw and I joined our XR friends in West Cornwall to present ourselves at Camborne Police Station, after carrying a coffin through the town and then making a very noisy protest outside. They refused to arrest us, but the mild satisfaction of getting the local press to write about the issue did little to alleviate the feelings of weary dismay that I felt, as all the restrictive powers that the Lords had managed to strike down were reintroduced in a new Public Order Bill announced in the Queen's Speech, shortly after our then Prime Minister Boris Johnson had been fined for breaking the law by partying during lockdown. This time, to the delight of *The Daily Mail*, the 'eco-hooligans' of Extinction Rebellion, Just Stop Oil and Insulate Britain were mentioned by name.[37]

'Could I suggest a more effective way would be legal action?' the prosecutor told Alice Clack, during the November 2022 trial of the Lambeth Bridge protesters. Perhaps groups and individuals suing governments, corporations and individuals for their harmful actions or inactions around climate and environmental issues is not exactly what the prosecutor had in mind. But if new legislation can be introduced to inhibit justice, why not use existing laws to further justice? According to the Grantham Institute, 'Climate change litigation continues to grow in importance year-on-year as a way of either advancing or delaying effective action on climate change.' Globally, the cumulative number of climate change-related litigation cases has more than doubled since 2015, and over 1,200 cases have been filed in the last eight years.[38]

We had gone this route ourselves at Greenham. Just prior to the cruise missiles being deployed, the US-based Center for Constitutional Rights sued the US government on behalf of a group of thirteen Greenham women and two congressional representatives. They argued that the deployment of the weapons created a substantial risk of nuclear war or of a nuclear accident, and thus broke international law. The congressional plaintiffs argued that deployment violated their constitutional right as members of Congress to declare war. In July 1984, the case was thrown out by the federal court in New York, where Rudy Giuliani acted for the defendants. The judge ruled that there were no judicially discoverable or manageable standards for resolving it, so he had no choice: 'The courts are simply incapable of determining the effect of the missile deployment on world peace. [...] The fact finding that would be necessary for a substantive decision is unmanageable and beyond the competence and expertise of the judiciary.'[39] This decision that the threat of Armageddon was beyond a jury's understanding was confirmed in the Court of Appeal, which declared that the issue of nuclear destruction was the prerogative of the elected branches of government.

But the simple action of going to court allowed us to publicise and discuss the issue in places which had previously been unreachable. I spent part of that winter of 1984 talking about nonviolence, the law and nuclear weapons to church congregations in rich suburbs in Pittsburgh and in rural farming communities in snowbound Ohio. So when I got a message on a WhatsApp chat in early 2022 saying that six Portuguese children and young people were taking legal action against thirty-three European governments, and the lawyers involved were looking for expert witnesses in mental health, I leapt at the chance to be involved. With the apparent support of the Global Legal Action Network (GLAN), the group were taking a case against these governments at the European Court of Human Rights (ECtHR), for failing to do their part to avert the climate catastrophe. Their goal: to get the ECtHR to make a legally binding decision requiring governments in Europe to take the urgent action needed to stop the climate crisis. This would require both deep cuts to emissions within their borders, and the taking of responsibility for emissions released overseas.

All the applicants had witnessed the devastating forest fires that had killed 100 people in Portugal in 2017, and realised this was a sign of things to come if global warming remained unchecked. In their testimony they noted the increasing restrictions on their own lives as temperatures rose in the summer, and the likelihood of more and more heat-related deaths. They talked of their own feelings of fear, worry and sadness about the future they were facing, as well as their witnessing of these feelings among their friends.

Consequently, they argued that climate change was interfering with their right to life, their right to respect for their private and family lives, and their right not to be discriminated against. By failing to urgently address the climate crisis, European governments were, without justification, forcing young people to face the burden of the worst effects in the future.

Interestingly, when the ECtHR asked European governments to respond to these charges, it took the unprecedented step of adding another issue: the youth applicants' right not to be subjected to 'inhuman or degrading treatment'. The fact that the Court raised the question itself, showed how the evidence and young people's testimony had affected the judges.[40]

I asked Ben Cooper KC, the lawyer who had reached out to me to be a witness, why this case mattered. He replied:

> It's the first time the European Court is applying European Convention rights to the climate emergency and addressing the impact of government inaction on individuals, on citizens, in the context of Articles 2, 3, and 8. Specifically, we are addressing the mental health impact of climate change under Articles 3 and 8, and that is why I contacted you and other mental health experts in this field.

A key point for him was that the European Convention on Human Rights (ECHR) is 'a living instrument'. The ECHR itself says that the Convention is supposed to be 'adaptable to new factual situations and to be applied differently as situations develop. So, it clearly has to be relevant to the situation where people are dying from all of the effects of climate change, and where people's mental health is suffering from climate change.'[41]

The Portuguese case is just one example of young people taking governments to court for failing to protect them from the impacts of the climate crisis. In spring 2021, the Federal Constitutional Court in Germany ruled that the German government should adjust its climate law because insufficient cuts in emissions violated young people's 'fundamental rights to a human future'.[42] That case built on an earlier landmark case in 2019, in which after years of litigation, the Supreme Court of the Netherlands ruled that the Dutch government's inadequate action on climate change violated a duty of care to its citizens and their human rights, as defined by Articles 2 and 8 of the ECHR.[43]

On the other side of the world, another lawyer, Melinda Janki, filed a constitutional claim against the government of Guyana in 2021 for acting unconstitutionally, and breaking its own laws, by permitting ExxonMobil's plans to extract 1 million barrels of oil a day off the coast. This would turn Guyana into ExxonMobil's single largest site of daily oil production within eight years.

And yet, thanks in part to Melinda, Guyana has some of the best environmental legislation in the world. She wrote the country's 1996 Environmental Protection Act, which defines the environment as including atmosphere and climate, and legislates for its management, conservation, protection and improvement, as well as the prevention or control of pollution, the assessment of the impact of economic development, and the sustainable use of natural resources. The Act also includes 'polluter pays' and 'precautionary' principles, which means companies are liable for the costs of cleaning up pollution, and the government is responsible for implementing measures to prevent environmental harm, even in the absence of 'full scientific certainty'.[44]

The Guyana constitution is one of the few in the world to state, in Article 36, that 'the well-being of the nation depends upon preserving clean air, fertile soils, pure water, and the rich diversity of plants, animals, and eco-systems'. Article 149J gives every Guyanese citizen 'the right to an environment that is not harmful to his or her health or well-being', and holds the State responsible for protecting the environment for the benefit of present and future generations.

Melinda drafted Article 36 and lobbied for Article 149. She already has the unique distinction of having beaten ExxonMobil twice in court, cutting their environmental permits down to five years, and getting rulings that they must produce an uncapped, unlimited parent company guarantee and indemnity to cover all the costs of an oil spill. In 2023 she received the prestigious Commonwealth Law Conference Rule of Law Award for her work to uphold the rule of law in Guyana. Her current case against the government is simple: 'The earth's atmosphere and oceans have been and continue to be polluted by the release and accumulation of greenhouse gases' resulting from 'the production, transportation, refining and use of fossil fuels. Our Constitution guarantees the right to a healthy environment,' Melinda explained to me, 'and we say that authorising the production of fossil fuels is contrary to the state's duty to protect the environment for present and future generations.'

Sitting in Friends House on Euston Road in London in February 2023, she told me that this was 'the biggest climate case in the world today. We are challenging a potential carbon bomb of about 4 to 5 gigatons.' If they are successful in arguing that oil production is not compatible with the right to a healthy environment, they will set a groundbreaking precedent that could be copied across the world. She then took me to task for using the term 'climate change': 'We've moved away from the cause to the symptom. The climate change movement drives me crazy because they're talking about a symptom and they're arguing about a symptom, when we need to get to the heart of the problem:

greenhouse gas pollution.' She saw this language shift as deliberately fostered by the fossil fuel industry to obscure the root problem, which is the pollutants produced from coal, oil and gas extraction. These cause not just global warming and climate change, but also air pollution and ocean acidification, hence the emphasis on the term 'greenhouse gas pollutants' in this case.

Melinda is an extraordinary woman. Born in Guyana, educated in England, she grew up in Guyana, Zambia and Trinidad. As a child she wanted to be a vet because of the deep connection she felt with animals, but decided that 'I was not clever enough to be a vet. So I chose law instead.' One thing that attracted her was 'exploring the different ways in which power operates, and the abuse and the misuse of power, and the way it comes at you disguised, and the assumptions that underlie that'.

Perhaps one of the reasons she felt able to challenge her own government and the fourth largest oil company in the world, is that she was familiar with how they operate. After completing her studies at University College London and Hertford College, Oxford, she went on to be one of the first non-white trainees at the City of London firm, Lovell, White & King (now Hogan Lovells), a corporate law firm 'otherwise known as "Lovely, White and Clean"'. She then worked as a lawyer for BP in London, until she realised that the world where people found high finance swaps and bonds 'sexy' was not for her.

She returned to Guyana as democratic government was being re-established in the 1990s, and did early work defending the rights of Indigenous people against mining companies, as well as drafting legislation to protect them. The current case is a Constitutional claim under Article 149J against the Attorney-General representing the State. This is in addition to five judicial reviews which aim to get the Environmental Protection Agency to follow the provisions of the Environmental Protection Act. The point of judicial review, Melinda explained, is that it is not about whether you are for or against oil, but whether you believe in the rule of law: 'The law is there and you go back to court to enforce it. We say, look, whether you are in favour of oil or whether you are against the oil, surely we can agree on one thing, which is that these guys have to obey our national law?'

This Constitutional case was filed on behalf of two plaintiffs: Dr Troy Thomas, a university Dean, and Quadad de Freitas, a young Wapichan tour guide and conservationist. Melinda was asking for 'Declarations of law, based on the facts that we put in front of the court.' As Dr Thomas put in his affidavit, the facts were that fossil fuel combustion was already harming the people of Guyana and 'the intensity of that harm will increase as fossil fuels continue to be burned'. Thomas quoted from ExxonMobil's own 1982 research, which concluded that 'mitigation of the "greenhouse effect" would require major reductions in fossil fuel combustion'.[45]

ExxonMobil disputed the fact that fossil fuels are a pollutant, arguing that it was a matter of opinion. Melinda contested this: 'There is no dispute of fact, fossil fuels are causing global warming and fossil fuels are causing ocean acidification.' The report from Exxon engineering showing that greenhouse gas pollution leads to an increase in temperature was included as evidence, as were statements made on oath by Darren Woods, the head of ExxonMobil, saying that ExxonMobil had long known about climate change.

As I listened to her, steely eyed, explaining the case to me, I was full of admiration for the patience, persistence and determination this kind of legal struggle requires. According to the Statista website, ExxonMobil is one of the top ten gas and oil producers and one of the biggest contributors to greenhouse gas emissions in the world, having released more than 40 billion metric tons of carbon dioxide equivalent since 1965. In the late 1970s, their own scientists warned of 'dramatic environmental effects before the year 2050'. Moreover, their predictions as to how much warming might occur were remarkably accurate. In 1982, while I was sitting outside that missile base, an internal memo from scientists stated that 'at the high end, some scientists suggest there could be considerable adverse impact including the flooding of some coastal land masses as a result of a rise in sea level due to melting of the Antarctic ice sheet'. Their response? To keep digging up more fossil fuels while 'overemphasizing uncertainties, denigrating climate models, mythologizing global cooling, feigning ignorance about the discernibility of human-caused warming, and staying silent about the possibility of stranded fossil fuel assets in a carbon-constrained world'.[46]

I asked Melinda how Guyanese people had responded to her campaign. Bharrat Jagdeo, then leader of the opposition, had denounced the oil deal: 'He said that our incompetent government trudged in there and stuck us with a deal that would harm us for decades. Now that he's vice president, he is defending the deal, refuses to do anything about it and is pushing the oil as hard as he can', she replied. Like the UN Secretary-General, she worried about the corrupting influence of oil, both in Guyana and globally: 'Oil is dangerous and it undermines the rule of law. Everywhere you go politicians listen to the fossil fuel sector rather than their citizens.' But people are stirring. Melinda told me about the daily protests outside a recent energy conference, 'in a country where public protest is not very popular':

> In the national press, there are articles criticising the government and the oil and criticising Exxon. So, there's tremendous anger against the oil and the exploitation of Guyana. People see it now as exploitation, they see it as abuse. It's only the government and the oil companies and their stooges who are desperately trying to convince people that this is somehow going to be good for Guyana.[47]

Legal actions are not just being taken against governments. An Indigenous Peruvian farmer, Luciano Lliuya, is suing the German

171

multinational energy company RWE for its share of the costs of preventing the glacial Lake Palcacocha from flooding his home-town of Huaraz. Glacial meltwater has swollen the lake, which now threatens 50,000 people in the town below as well as Lliuya's own land and livelihood.[48] In 2021, a Dutch court ordered Shell to cut emissions by a net 45 per cent, both from its own opera-tions and from the use of the oil it produces, in response to a class-action lawsuit by Dutch NGOs and citizens. Moreover, Shell was obliged to begin even while appealing the decision.[49]

New targets for climate litigation spring up every day. In July 2022, ClientEarth, a charity that specialises in using the law to protect the planet, along with Friends of the Earth, the Good Law Project, and environmental campaigner Jo Wheatley, suc-cessfully sued the British government for breaching its own Climate Change Act 2008, because the net zero strategy fails to show how the government will meet the legally binding carbon budgets established by the act. The government were given eight months to put it right.[50]

'Actually, when you are talking about law and climate break-down the biggest story is not the cases brought by people defending the planet, it's the hundred law firms who have sup-ported almost 2 trillion dollars' worth of fossil fuel transactions.' This comes from former barrister Tim Crosland, who biked over to meet me in January 2023 at a café on the South Bank, appro-priately enough placed between the Houses of Parliament and the City of London. He regarded much of the excitement around climate litigation as a form of greenwashing:

a nice, big lie for the legal profession because it portrays them as sort of coming to the rescue. What hardly gets talked about is the legal process that has meant $100 billion of public funds go into fossil fuel companies to compensate them for the crappy climate policies which governments have hardly actually enacted, while they're also making gazillions from the war in Ukraine.

He was referring to the Energy Charter Treaty which allows corporations to sue governments for climate policies that might erode their future profits, and thus disincentivises governments from enacting such policies. Seven EU countries already planned to withdraw from the treaty because it undermined the Paris Agreement of 2015. The British government, unsurprisingly, was procrastinating.[51]

Tim has spent much of his legal career riding to the rescue. After doing a masters at Utrecht in international human rights and environmental law, he worked as a human rights lawyer, first for Kent Police and then as head of the legal department at the National Criminal Intelligence Service. Working there helped him develop an idea of 'law as resistance' where 'using the language of the people within the system and using the ideas that they claim to be important to them' exposes the internal contradictions. It was also where, on a work trip to Nigeria, he reconnected with environmental issues. He learnt how some 20 million refugees were being driven into Northern Nigeria because their primary water source, Lake Chad, had lost 80 per cent of its water volume since the 1960s, largely because of climate change, 'and I start to feel the irony of being sent out by the Foreign and Commonwealth Office to Nigeria, to talk about legal frameworks and human rights, when we're driving so much of this insecurity with our policies in the UK'.

He left government work in 2015, and after immersing himself in environmental and climate science went to COP21 in Paris as a legal adviser to the Small Island Developing States,[52] employed by a legal charity funded by the then Department for International Development (DFID). It was at this meeting that the famous commitment to keep global warming below 1.5°C was made. What I had not realised before meeting Tim was that initially the US, Canada and Europe told the Small Island Developing States and others: 'We will give you the 1.5°C in the text of the agreement, but only if you waive your rights to sue us.'

Fortunately, there were many parties, including Tim, who were determined to block this. But DFID, through his charity bosses, told him to stay out of the fight: 'I went off-piste with this and they sacked me.' But the right to sue was not waived, and Tim learnt a crucial lesson—that governments would promise anything as long as they were not held accountable: 'Because if we are held accountable, the scale of the damages is so vast, it will end up being a massive transfer of wealth back to the people we stole it from, and that we do not want.'

This was the genesis of 'Plan B', the legal charity founded in 2016 by Tim and others, 'to hold to account those in power who knowingly take action inconsistent with the 1.5°C temperature goal of the Paris Agreement—the lifeline for humanity and for so much other life'.[53] In 2020, their case against Heathrow Airport did exactly that when Plan B persuaded the UK Court of Appeal to rule that government plans to build a third runway at Heathrow were illegal because they failed to take the Paris Agreement into account.

Unfortunately, the judgment did not stand. Heathrow Airport Ltd appealed, and at the end of 2020 the Supreme Court ruled in their favour, saying the Paris Agreement had been taken into account, so the runway could go ahead. Tim broke the embargo on the judgment the day before it was due to be made public, in order to draw attention to what Plan B regarded as both the government and the Supreme Court's suppression of evidence that the Heathrow expansion was inconsistent with the Paris Agreement threshold of 1.5°C. Explaining this unprecedented act of civil disobedience, he wrote in a statement on the Plan B website:

> This will be treated as a 'contempt of court' and I am ready to face the consequences. I have no choice but to protest the deep immorality of the Court's ruling. [...] I have been a lawyer for 25 years [...] I have deep respect for the rule of law and the vital role of the judicial

system in holding power to account. That is why it is a duty to protest a decision that so gravely betrays that purpose. I take this act of protest for the sake of my two children and in memory of all those who have lost their lives on the frontline of the climate crisis, in the UK and around the world.[54]

He was found in contempt and fined, and has since resigned as a barrister, prior to being disbarred. Resignation was an important step for him: 'If we have a legal system that's deeply invested in the carbon economy and has been for centuries, that influences the way they interpret the law.' So, for example, 'although the right to life is an existing part of our legal system and should mean governments must take practical measures to protect one, we know the courts in this country will not choose to interpret it in that way.'

One reference point for Tim is Germany in the 1930s and 1940s. His maternal grandmother was a German Jew, and the role German lawyers chose to play at that time and their subsequent trials at Nuremberg for the part they played in crimes against humanity presents the 'great moral paradigm. [...] If you see what's happening now as a crime against humanity, and all life on Earth, it's just a matter of time before that reality somehow catches up with us, one way or another.' He wanted to have no part in this: 'I don't respect the authority of these courts, while they're acting in this way. And the logic of that is I must step away from a position that requires me to respect the authority of the Courts.' He was not asking others to do the same, but 'just to think about what you know, what the law is really doing and how we can influence that'.[55]

One method of influence was to join with 250 other legal professionals in writing an open letter calling on lawyers 'to profoundly change their practices by engaging in climate conscious lawyering, warning that breaching the 1.5°C temperature goal established by the Paris Agreement on Climate Change risks mass loss of life and threatens the conditions for a stable civilisation, including the rule of law.'[56]

After talking to Tim, I walked along the Thames Embankment to Southwark Crown Court, where those seven women who broke the glass at Barclays Bank in 2021 were being sentenced. In contrast to the Shell Six, the jury found them guilty of causing £100,000 worth of damage, and they all expected to go to prison. I joined some hundred women who had gathered in solidarity, many dressed as suffragettes, to listen to Helen Pankhurst, granddaughter of Sylvia, remind us: 'It takes courage to challenge the status quo, to risk prison for a cause, and to speak truth to power.'

Judge Alexander Milne KC told the women that he regarded their action as a 'publicity seeking gimmick', but that it was mitigated by the care they had taken to avoid harm to anyone, and their 'sincerely held beliefs'. They were all given two-year unconditional suspended sentences and £500 fines. Perhaps Judge Milne had read the lawyers' open letter.[57]

CHANGING IDEAS OF AUTHORITY, NOVEMBER 1984[57]

We are lying 10 yards from the silos. An owl swoops over us, wings caught in the headlights of a passing car. We have crawled, run, walked across the runway and through woods to get here and are now waiting for the evening rush of traffic to diminish before running through the last brightly lit area, in an attempt to get over the inner fence. Groups of women have been doing this every night. On one occasion 21 hijacked an airforce bus, drove up to the inner silo fence and started cutting before being arrested. The actions continue in spite of the fact that the camp is under the worst pressure for months. Evictions occur sometimes twice a day—they aren't pleasant. The bailiffs on one occasion emptied the Elsan on the ground before taking it. The water standpipe has been taken and the Water Board, having decided we are 'illegal', has refused to replace it.

I am soaked to the skin and covered in mud, my coat was lost crawling through the first lot of wire. There are three rabbits hopping about in front of my nose. The traffic seems endless. Why am I doing this? The media aren't interested, haven't been for months. Am I wasting my time?

But how do we define effective action? Greenham has long stopped being a publicity exercise. The actions over the last year—the occupation of the air traffic control tower, the perpetual incursions and blockades, the removal of sections of fence and documents from inside the base—have changed from mass symbolic set pieces to a form of perpetual resistance; the nearest non-violence can get to guerrilla warfare. They have not lost their symbolic value. On Nagasaki Day the women stripped naked, covered the ground and themselves with ashes and then blockaded. But they have far more to do with preventing those who run the base operating efficiently than with publicity seeking. In this it would appear that we are being effective. A friend visiting the USAF Commander recently remarked that his desk was covered in newspaper cuttings about the camp. The Defence Committee Hearing at Greenham on physical security last May was entirely preoccupied with the camp.

Not that their discussion showed much understanding of our psychology. Defence Minister Keith Speed suggested that an extremely dull 10ft-high wall with TV surveillance on top should be built. Having nothing to watch we would get bored and go away. Mr Ward (head of S.I.) did point out that this would appear to be walling in the Americans and thus provide the peace camp with a sort of victory. 'So far as is possible we would like Greenham not to become so much like a prison camp that it resembled no other American Airforce base in this country.'

Meanwhile the tracking of the cruise convoy has become increasingly efficient. So far there has been no secret dispersal at all, while on the last exercise five peace activists in a van actually joined the convoy for a period. Wing Commander March admitted to Evelyn Parker of Newbury Campaign Against Missiles that the convoys had to travel at 'indecent haste' whereas 'under normal circumstances' (!) this would not be necessary.

The MoD spent £8.75 million on overtime last year—£6 million of that on demonstrations. Far from occupying a few troublemakers the ten-day action in September demonstrated the power of the camp to attract thousands of new women from as far away as New Zealand and the USA, and as Liz pointed out, 'Women who live here any length of time simply don't go back to being housewives.' Greenham for many is their first contact with organised violence in the form of dogs, army, police, soldiers and barbed wire—it is unforgettable. 'They see women being arrested on quite trivial charges while significant laws such as the Genocide Act are ignored. It completely changes your attitude to authority.'

Few women pass through the camp without deciding that cruise missiles are just one particular and frightening manifestation of our complete lack of self-determination. So that it seems completely natural to move from here to campaigns against Violence against Women; support for Nicaragua, Namibia, or miners' picket lines. Some sections of the peace movement see all this diversity as diluting the real issue of nuclear weapons. And it can't be denied that, while we find it easy to see and explain the connections between these issues, working in more than one place at once is difficult.

Sarah Hipperson, writing in a recent issue of *Women for Life on Earth* described her disappointment at the fact that thousands of women don't come to the camp to obstruct the cruise convoys on their return. 'I remember raging bitterly that Greenham women may be everywhere but they certainly aren't on Greenham Common while this "beast" is out travelling through our country preparing for mass murder ... If a thousand women came to Greenham to "picket" think how stretched the police would be looking after all the pickets in the country. Indeed if this was done once a week that would be a practical as well as political promotion of nonviolence to strengthen and solidify it.' However, connections work both ways and making them has brought other women to Greenham—black women, miners' wives.

SORRY FOR THE INCONVENIENCE

The strangest thing is that I am not scared. Gathering with other women outside the fence in the dusk I was terrified: images of myself shot, caught on barbed wire, beaten up by soldiers or bitten by Alsatian dogs, flickering on the edge of my mind. They have cleared now and lying in the mud beside Hera and Skippy, women I love and trust, I feel quite safe, intent only on how we might get over that fence, and what has happened to the other women. Those who worry that the camp has lost its sense of direction by using its energy to maintain a women's community, rather than challenge cruise, miss the point that the only possibility of an effective challenge comes out of creating that community.

That's not to say we have Utopia. I can't think of a more insane place to try and create it than on a public highway outside a barbed wire fence with no shelter worth its name. We have had bitter quarrels over the use of money, the ability to act as we choose without limiting others' freedom to do likewise. We have seen women come here mad with distress, and have failed utterly to help them, and then ourselves left in disgust … and come back again, because somehow the centre holds—a shared belief that even if we can't live up to our own ideals the effort is worth it.

Sometimes it works—over the ten days' action, workshops were held, shit pits dug, litter collected and countless actions taken with no sense of a central leadership. When the evictions began in earnest again several weeks ago, women appeared miraculously with hot food; water is dropped off by total strangers at least twice a day. There isn't a short cut. If getting rid of cruise means abandoning the attempt to look after each other it isn't worth doing.

The traffic has lessened, it seems quiet. We get up and start to run towards the wire. A soldier appears from behind an observation box. Another appears 100 yards away down the fence. 'Stop running', he yells. We put our hands up, smiling and walk towards him. He looks incredibly young and scared. Fifteen soldiers anxiously gather around seven women and push us into a Land Rover. We're driven to the administration block and taken out one by one to be searched. It's interesting to see how neatly the services mesh together. Cathy, plump and blond in silky blue bomber jacket (USAF), is acting tough with Skippy—'Siddown or I'll sidya down'. Sarah is pacing up and down while a nervous RAF man follows her with an exasperated expression on his face. 'Sit down!' he says in desperation. 'You're working for death,' she replies, 'and I'm not doing anything you say!'

Cathy calls for help and a large MoD policeman arrives. 'Look Miss Green, you know the form, sit down, please.' (There are no chairs and the only reason I am sitting on a very cold floor is to write notes.) 'You know the form—you're preparing for murder! Why have you made that woman sit alone?' 'Because I fancy her and I don't fancy you!' Another MoD officer arrives, glances at her and at the series of open doors— 'Can I have those doors locked please!' Last time Sarah was here she removed lecture notes discussing the operational deficiencies in the running of the base and its preparedness for chemical and biological warfare. 'Didn't I tell ya it was goin' to be one of those nights?' Cathy remarks to no one in particular. 'It will be "one of those nights" every night until you get rid of cruise', Sarah retorts.

There's much discussion in the peace movement as a whole on the need for new ways of thinking; a growing realisation that fear of the effects of nuclear weapons is not the best basis for action. Indeed there's something paradoxical about first terrifying people and then expecting them to be brave enough to challenge a government armed with that kind of might.

So we are discussing realistic alternatives. The only problem for me is that many of the new alternatives seem remarkedly like the old. I am no less terrified by the prospect of conventional warfare, however defensive. While we remain in a society whose organising principles are violence and exploitation, the lesser of two evils is still evil. Yet pacifism is not discussed; it is seen as a form of 'weakness' or 'giving in' and everyone is still frightened. If Greenham stands for anything it is to challenge that. 'Non-violent action', Liz remarked to me one day, 'is simply a way of overcoming fear of one sort or another'. I see the continuing existence of the peace camp as a constant reminder of the possibility of non-violence as a creative struggle that doesn't have to end in defeat. For three years a small bunch of women have successfully held onto a patch of land in the face of a government with all the coercive powers of the State at its disposal and at the same time they have created a network and philosophy of personal responsibility that stretches across the world.

They haven't charged us and we are released in twos, walking past the security box and through a gate that divides two realities—ours which says that conflicts can be resolved without violence and theirs, where the presence of violence prevents anything being resolved at all.

6

HOW TO PREVENT HARM

After some four years of climate activism, I find myself asking the same questions we asked in 1984. Should we be focused on simply alerting the public to the dangers they face from the climate and ecological emergencies, or are there other nonviolent ways to intervene more directly to stop the damage?

Direct nonviolent intervention can be profoundly satisfying. And it can be effective. The cruise missiles' invulnerability to Soviet attack depended on the fact that at times of crisis they would be dispersed to secret locations in the surrounding countryside. As early as January 1984 *Jane's Defence Weekly* reported that 'training exercises had been shelved because of political problems caused by the continued presence of the peace protestors'.[1] In 1985, General Charles L. Donnelly Jr, Commander in Chief of the US Air Force (USAF) in Europe, admitted that protesters had reduced by 25 per cent the frequency with which convoys were able to move outside Greenham.[2]

In spring 1984, no doubt in part because the government was concerned at the missiles being trapped inside the base, the authorities began serious attempts to 'finally' evict the women.

They failed. Destroying shelters, shredding sleeping bags and extinguishing fires simply taught us to become yet more flexible and mobile, and learn to live in the open. It was a much harsher existence. The press had begun a concerted campaign against us, and local vigilantes also attacked the camp. But sustaining our presence had become part of the action. Greenham was no longer about alerting the public to the presence of cruise missiles and the dangers of nuclear war. As described above, it was about persistent nonviolent confrontation with the nuclear state, a sort of nonviolent guerrilla warfare with humour. This took the form of frequent incursions into the base, including penetration of the high security areas; the removal and publication of secret documents; occupation of the air traffic control tower; commandeering of vehicles; painting and roller-skating and bicycling down the runway: an endless series of actions that made a mockery of military security. The element of ridicule was an important part, as were growing relationships of mutual understanding, if not always respect, between some base personnel, the police and the women.

Sometimes base personnel responded with violence. On one occasion, when I and two friends had climbed on top of a fuel pump inside the base, a young USAF soldier climbed up after me and began to push and shove me, threatening to throw me the 15 feet to the ground. When I asked to be allowed to climb down, he let me begin, then brought his booted feet down heavily on both hands as I clung to the pipework, grinding them in and making it impossible to move. He only moved his feet when his colleagues arrived and shouted at me to come down.

But harassment and aggression did not stop us. Women had extended their area of operations to all the surrounding nuclear facilities: entering the Atomic Weapons Research Establishment (AWRE) at Aldermaston, the bomb stores at Welford, and climbing into F-111 aircraft at Upper Heyford. Over the next three years, Greenham women, in combination with a growing network of nonviolent resistance called 'Cruisewatch', became expert at harassing and disrupting the supposedly secret dispersal exercises. This involved closely monitoring the base at all times and, as soon as an exercise began and a missile convoy left the base, tracking and following it. Different groups would take different actions, some blockading the road and others throwing paintballs to mark the convoy and show those in charge that the convoy had been seen.

I remained involved even after starting work as a trainee GP in Cornwall in 1985, driving up to support the camp and join in Cruisewatch actions when I could. I remember the astonishing feeling of empowerment when, after doing night watch with a handheld radio in my beaten-up Mini Traveller, myself and Lisa, a fellow Greenham woman, were able to closely follow a convoy of fourteen vehicles and four missile launchers down the A34 and alert the wider network. On another occasion I was arrested, along with other friends, after following the convoy onto Salisbury Plain and handing out leaflets to soldiers; but after a night in the police station on the Plain we were once again released without charge.

That decision not to charge us was in marked contrast to how peace activists taking similar action had been treated in the past. In December 1961, six nonviolent anti-nuclear activists planned and organised a blockade and mass trespass of the NATO air base at RAF Wethersfield in Essex, because of its role in hosting preparations for nuclear war. They were all leading members of the Committee of 100, part of the anti-nuclear movement that was committed to nonviolent civil disobedience that directly targeted nuclear missile bases and government institutions. The Committee had already organised large and impressive sit-down demonstrations in Scotland and London, in which more than 1,000 had been arrested and charged with public order offences.

On this occasion, the six were charged and successfully prosecuted for conspiring to commit a breach of the Official Secrets Act (OSA) by entering the air base for a 'purpose prejudicial to the safety or interests of the State and second, conspiring to incite others to do likewise'.[3] The five men received prison sentences of eighteen months; the one woman, twelve months. And even though some 5,000 people demonstrated at Wethersfield, with more than 800 arrests, many argue that these punitive sentences discouraged further civil disobedience at this time.[4]

Mrs Thatcher, in contrast, adopted a more dovish approach during her second term of office.[5] Concern about public feelings on the nuclear issue, and sensitivity to a growing debate on civil liberties, is probably what restrained the government from using the full force of the law against Greenham women. Therefore although on many occasions women were charged under the OSA, they were never prosecuted. The government was no doubt wary of the likely impact of an Old Bailey trial, and conscious that a jury might not necessarily convict. So charges were manipulated to confine women to an unsympathetic magistrates' court.[6] By 1986, Greenham arrests had topped 4,000.[7]

Cruisewatch and the women's camp between them ensured that not a single dispersal exercise took place in secret, and exposed the myth of invulnerability to detection and sabotage. After all, if a bunch of peace activists with dysfunctional radios and no mobile phones could track a nuclear weapons convoy and cover it with paintballs, what could the Soviets do?

In 1987, Gorbachev and Reagan signed the INF Treaty, which meant that intermediate nuclear forces in both halves of Europe—SS-20s and cruise and Pershing missiles—would be removed over a three-year period. The negotiators acknowledged that pressure from the peace movement had helped in achieving this goal. Nonviolent protests across the globe had resulted in the first treaty ever to reduce the number of nuclear weapons.

In his seminal and influential three-volume work, *The Politics of Nonviolent Action* (1973), political scientist Gene Sharp describes some forty-one methods of nonviolent intervention. They differ from protest, in that the aim is to have a direct impact on behaviours, policies or institutions.[8] The interventions might be negative, through disrupting or preventing activities from occurring, as we hoped to achieve by interrupting the launchers and base activities at Greenham; or they might be positive, leading to the establishment of new policies and practices.

One of the most famous positive interventions were the Freedom Riders, whom I mentioned in Chapter 2. Their courageous, nonviolent bus rides into the Southern United States helped to achieve the desegregation of interstate transport. They inspired generations of nonviolent activists, including today's climate activists. What I had not realised until reading Sharp was that there were earlier freedom riders in the nineteenth century, campaigning against segregated transport on steamboats, stagecoaches, streetcars and railroads, in many US states. As in the later campaigns, white riders might ride in Black areas or Black riders ride in white compartments, often facing significant abuse. In Louisville in Kentucky, for example, when local streetcars refused to obey a federal ruling to desegregate, newly freed Black slaves began a 'ride in' campaign. This provoked white violence against the riders, while the drivers often abandoned the cars. Moderates in Kentucky protested, court cases followed, national attention grew and the federal government backed the Black people. As a result, all city transport in Louisville permanently abandoned segregation in 1871.[9]

A similar dynamic would play out in the most famous freedom ride campaigns of 1961, sponsored by the Committee of Racial Equality, CORE. As with the earlier Journey of Reconciliation, organised by George Houser and Bayard Rustin in 1947,[10] they were actually upholding the law, enacting their constitutional right to non-segregated travel on interstate transport, granted by a Supreme Court ruling in 1946. Over a six-month period, there were at least a dozen long-distance rides involving more than 1,000 people. The extreme violence meted out to both Black and white riders in the Southern states in the form of physical beatings, bus burnings and attacks by mobs, including the Ku Klux Klan, along with the failure of local law and order officers to protect the riders, drew the support of other organisations, and attracted international attention and outrage. This courageous preparedness to accept violence without using it oneself, exemplifies what Sharp and others have called the jiu-jitsu effect of nonviolence, where the protesters refusal to respond to violence in kind, even when facing severe repression, throws those using it off balance, and the sacrifice involved attracts wider support.[11] In November 1961, the Interstate Commerce Commission finally enacted a ruling that allowed passengers on interstate buses and trains to sit where they wished, regardless of race. All facilities inside terminals were similarly desegregated and white and coloured signs removed.[12]

Sharp's writings also introduced me to other early nonviolent interventionists such as those wanting to protect runaway slaves in Northern US states from the 1850 Fugitive Slave Law, which required private citizens to assist in their capture. Abolitionists called upon their fellow citizens to 'spit' upon the law, while pastors reminded their congregations of the duty to disobey unjust laws. Pacifist pastor Samuel J. May told his Unitarian congregation to 'trample this infamous law under foot, be the consequences what they may [...] It is not for you to choose, whether you will or not obey such a law as this. You are as much under obligation not to obey it, as you are not to lie, steal, or commit murder.'[13]

Others recommended what Sharp labels as nonviolent harassment: the watching and public condemnation of known slave hunters paid to kidnap and return slaves to the South. This would be done by posting handbills with their names in public places, and persuading landlords to refuse them lodging or entertainment. If this did not happen: 'Whenever they go out two resolute unarmed men should follow each of them wherever he goes, pointing him out from time to time with the word Slave Hunter.' The method was effective. Sharp gives the example of a Miss Wilson from Maryland who came to Philadelphia to find her runaway slave. When she was identified and named on large posters around town which said 'Beware of Slave Catchers' and exhorted people to hide the named slave, she abandoned the hunt and went home.[14]

For my generation, children born in the 1950s, in the shadow of the Second World War, the nonviolent interventions that caught our imagination at school were the stories about the protection and rescue of Jews and other fugitives. This was one of the many significant methods of nonviolent resistance to the Nazis. Such stories had particular resonance for me, as my maternal grandfather was an Austro-Hungarian Jew. He had left Vienna in the 1920s to work for the London branch of his business, and was followed by other family members. But not all: my great Aunt Francie threw herself out of a bedroom window in 1938, when Hitler entered Austria; and other family members who remained in Austria did not survive. But thanks to the astonishing courage of rescuers, many did.

One of the most famous examples is the village of Le Chambon-sur-Lignon, where the local Protestant pacifist Pastor André Trocmé and his wife Magda inspired and organised their fellow villagers and others in the surrounding community to take in Jews and others fleeing the Vichy government and the Nazis. During the four years of occupation, some 5,000 people were rescued and hidden in plain sight in local schools, homes and farms. Some were escorted over the border to Switzerland. Pastor

André was imprisoned for a period and others paid with their lives, but it remains one of the most powerful examples of effective nonviolent resistance to dictatorship.[15] It is estimated that possibly as many as a million Jews across Europe were saved by rescuers taking similar actions.[16]

When looking at nonviolent interventions, it is interesting to note how often healthcare workers are involved. Perhaps it is not surprising. The healthcare professions tend to attract those who want to 'do something' to help others. We are mostly 'interventionist' by nature. All medical students are taught about Dr John Snow and his determination to discover the source of a cholera epidemic in Soho in London in 1854. The disease was thought at that time to be spread by infection through the air. It was Snow's meticulous tracking and mapping of cases of infection which showed that all of those with cholera had used water from the same source: the Broad Street pump; whereas those who used other sources of water were cholera-free. Snow persuaded the local authorities to remove the handle of the pump—a simple intervention that prevented people from using this water source and brought the outbreak to an end. He then went on to show how sewage contamination of water was the source of the infection.[17]

Dr Snow was acting within the law. Not so Dr Alex Woldak, the director of a drug and alcohol service in Sydney, Australia, who in 1986 recognised that 'rampant HIV was spreading among and from people who inject drugs to the general community' and knew he had to act. His intervention took the form of establishing and funding a pilot needle and syringe exchange, illegal at the time. The New South Wales Department of Health told him to close it or face legal action. He refused to do so, and the police, when shown the effectiveness of his programme, did not press charges. By 1987, needle and syringe exchange programmes had been legalised by the NSW government and were being run from pharmacies and community clinics. Later they went nationwide. The result: injecting drug users in Australia have one of the lowest HIV infection rates in the world.[18]

In the same year that Dr Woldak was setting up illegal needle exchanges in Australia, 420 healthcare professionals on the other side of the world were protesting in the Nevada desert, on the same day the federal government was conducting a nuclear test. Of these, 139 crossed the line onto the United States Department of Energy underground Nuclear Test Site and were arrested. The American Public Health Association actually made the demonstration at the site part of their annual meeting, held in the nearby Las Vegas Hilton. They advertised it in the official programme, sold tickets outside all the main conference meeting rooms, and laid on buses to drive participants to the Test Site. The aim was to push the US to join the unilateral moratorium on nuclear testing that the Soviet Union had begun fourteen months earlier.[19]

All these actions are in keeping with the duties of a doctor to 'protect and promote the health of patients and the public', as my colleagues had told Judge Robinson when on trial for sitting on Lambeth Bridge at the City of London Magistrates' Court in November 2022, and as the General Medical Council states in its description of Good Medical Practice. This also requires doctors to 'make the care of your patient your first concern' and 'take prompt action if you think that patient safety, dignity or comfort is being compromised'.[20]

The government's own guidance on Climate and Health states that the UK is 'particularly at risk of drought, flooding and extreme weather events, all of which threaten the water, food, infrastructure and supply systems we depend on'. It then goes on to say that 'Health and care professionals should recognise the climate crisis as a health crisis, and therefore climate action as a core part of their professional responsibilities.'[21] Thus the interventions described above, whether trying to prevent the spread of infectious diseases, end the danger of nuclear annihilation or address the climate and ecological emergencies, could clearly be regarded as part of a doctor's duties, even though they contradict another professional requirement, which is to act 'within the law'.

Throughout the 1980s I was repeatedly told by my bosses, colleagues and family that my nonviolent law-breaking, including a brief spell in prison, would mean the end of my medical career. Yet although my criminal record was always fully declared and detailed on every job application, fortunately I continued to be employed and remain on the Medical Register. To date, as far as I know, no medical colleague of mine has been struck off for engaging in civil disobedience. In 2020, a *Lancet* editorial argued that 'institutions and the public should firmly support the right of health professionals to participate in climate action and to be protected from censure by medical colleges or licensing bodies if such civil action results in arrest or charge'.[22]

My own choice today to take supportive rather than arrestable roles in current actions in Britain comes from my wish to continue to engage in another form of intervention: providing medical assistance to those forced to flee from conflict and disaster.

I have always been someone who, when faced with a problem, needs to be right up against it. Cutting down a fence and climbing in and out of a nuclear missile base which housed preparations for genocide was one way. But in 1991, war arrived in the country that was then my home, Slovenia, which declared independence from the former Yugoslavia in June of that year. The following summer there were Bosnian refugees living just down the road, so I started to work with them. With the support of my extraordinarily tolerant boss, Professor Ian Goodyer, I managed somehow to combine a training in child and adolescent psychiatry in Cambridge, with commuting to the Balkans to carry out relief work and research.

In the following decade I worked for a large medical NGO. My job was to go into various emergencies, assess needs and establish mental health programmes. In almost all of them I witnessed at first hand the degree to which the climate and ecological crises either directly precipitated or exacerbated what was happening. I worked with Kurdish children in Northern Iraq who had already lived through five fossil fuel-fed wars, before the UK's illegal intervention in 2003. In Chad I worked with starving refugees from Darfur. They were farmers driven out of their country by land-hungry pastoralists, themselves endangered by years of desertification as the Sahara expanded to the south.[23] Working in devastated New Orleans after Hurricane Katrina in 2005, taught me that there is no such thing as a natural disaster. Poor communities and people of colour were three times more likely than wealthy, predominantly white communities to lose their homes and become displaced by the storm.[24] As Juliette Landphair states in her paper 'The Forgotten People of New Orleans', 'the Lower Ninth Ward came to represent the convergence of destructive forces on a society: the hurricane; the geographical vulnerability of New Orleans; government neglect; and urban poverty and racial polarization'.[25] In Haiti I witnessed how the serial hurricanes and mass deforestation limited the hopes of 'building back better'—that much-loved slogan of relief workers across the world.[26]

And then the disasters came home to us. In 2015, the steady, ever-growing flow of those trying to find refuge in Europe— from conflict, disaster and impoverishment—escalated dramatically, as people fled the war in Syria. That long-running conflict also exemplifies the exacerbating role of the climate crises. Between 2006 and 2010, one of the worst multi-year droughts in history extended the desert into large areas of the country, destroying agriculture and driving more than 2 million people off the land and into cities, where they added to the existing economic instability and civil strife created by the dictatorship.[27]

And as thousands made perilous journeys across the Mediterranean and found their ways blocked at various choke points, such as the English Channel, the migrants began their own forms of nonviolent intervention: site occupations. With the help of volunteers from local communities and from abroad, they created informal settlements across Europe. Asmamaw and I were among the many trying to offer support, in Greece, Italy and France. These settlements, built from rubbish and recycled and donated materials, and the communities, were astonishing. Of course, as everywhere, there were fights, petty crime and abuse. But what we also witnessed was the extraordinary capacity to create supportive communities out of nothing; to produce art, music and theatre; and to share and care for strangers. In the Jungle in Calais in early 2016, there were volunteers and migrants providing healthcare, legal advice, and education for children and adults, and running kitchens. In the evenings I attended regular meetings where volunteers and migrants shared information, and discussed and organised together how resources should be distributed and conflicts resolved. It was an egalitarian, self-organising system of the kind I had not experienced since living at Greenham women's peace camp.

And when the authorities in different countries began evictions and site clearances, I witnessed other forms of nonviolent resistance, what Gene Sharp labels 'psychological interventions', designed to change behaviour through 'moral pressure'. When the French authorities closed down half the Jungle camp in Calais in March 2016, nine Iranians sewed their mouths shut and went on hunger strike. They ended the strike when the authorities promised to improve conditions in the other half of the camp. Another Syrian friend, former law student Housam Jackl, stood every day in silent protest on the closed Greek-Macedonian border.[28]

Over the years, as with other forms of protest, there has been a steady criminalisation of those trying to assist migrants, turning compassionate actions and interventions into civil disobedience by default. In 2017 in Ventimiglia on the Italian–French border, three volunteers were arrested simply for distributing food bags to migrants, an activity which had been banned by the mayor two years earlier. More disturbing is the criminalisation of those trying to save migrant lives at sea. In the last decade, more than 23,000 people have drowned trying to cross the Mediterranean. The rescue at sea of those in danger of being lost, regardless of the nationality of the victims, is mandated and protected by both the 1982 UN Convention on the Law of the Sea (UNCLOS) and the 1979 International Convention on Maritime Search and Rescue (SAR Convention).[29] Yet the crew of the *Iuventa*, an NGO search and rescue ship which had saved some 14,000 people, operating legally in cooperation with various coast-guards, were trapped in Trapani in Italy in a sting operation in 2017.[30] The ship was confiscated and its ten crew members arrested and charged with 'aiding and abetting illegal migration'. At the time of writing, charges had been dropped against six crew members, but four remain in process and the boat has been destroyed.

When migrants themselves intervene to try and save lives, responses have been even more punitive, as *Iuventa* scrupulously documents on its website. Migrants are 'criminalized for having helped other migrants in need, or for having been forced to skipper the boats bearing them across the sea to Europe. Greek courts routinely sentence migrant skippers to an average of 44 years in prison and a 370,000-euro fine on the basis of 30-minute hearings.'[31]

One of the most dramatic nonviolent interventions on behalf of migrants actually occurred in the UK in March 2017. Nine women and six men cut a hole in the perimeter fence at Stansted Airport and then surrounded a Titan Airways Boeing 767. Some locked themselves together in front of the plane; others locked themselves to a tripod near one of the wings. The plane had been secretly chartered by the Home Office to deport sixty vulnerable migrants back to Nigeria, Ghana and Sierra Leone. Flights were diverted and it took some hours to free them. They were arrested and initially charged with aggravated trespass, facing a maximum three-month custodial sentence. However, the charge was changed to putting the safety of the airport and passengers at risk and causing serious disruption to international air travel. These are terrorism-related charges that can result in life imprisonment under the Aviation and Maritime Security Act 1990.[32]

The defendants argued that far from endangering anyone, they needed to act because they were trying to protect those on the flight from harm. On their journey to Stansted, before the demonstration, they had read out testimonies to each other, from some of those they were trying to save:

> One was from a Nigerian woman, a lesbian who was being held in Yarl's Wood immigration detention centre. She'd been forced into a marriage, had kids and then fled Nigeria for the UK. Her children were staying with her sister and she was sending back money to help support them. Days before she was due to be deported, her ex-husband contacted her to say that he knew she was being returned, and that he'd be waiting there to kill her.[33]

After a ten-week trial, the jury found the defendants guilty. However, when Judge Morgan came to sentencing, he acknowledged that all the defendants were motivated by 'genuine reasons' and that they 'didn't have a grievous intent as some may do who commit this type of crime', and he gave them all non-custodial sentences.[34] Their convictions and sentences were quashed on appeal in January 2021, because, as the judgment stated, 'their conduct did not satisfy the various elements of the offence' and 'there was in truth, no case to answer'. It is to be hoped that those considering bringing terrorism charges against nonviolent activists will now be more hesitant: and the action did protect some of those at risk. In 2021, eleven of those due to be deported remained in Britain, including one identified as a victim of trafficking who was granted asylum.[35]

The environmental movement has a long history of effective nonviolent intervention. I learnt at school how, in April 1932, a group of working-class young men (our teachers failed to tell us they were young communists) organised a mass trespass on Kinder Scout in the Peak District, to reclaim the right to roam freely in the countryside. Although once common land, like so much of Britain, it had been enclosed in the seventeenth century and was owned by the Duke of Devonshire. Neither the Derbyshire police, nor the Duke's gamekeepers could prevent some 600 young people walking up the hill, but they did arrest five. They were tried, found guilty of Breach of the Peace and Riot and sentenced to up to six months' hard labour. It was this repressive response that led to a public outcry and increased interest in the issue. The Ramblers' Association had been negotiating for access over the previous six decades with only moderate effect. That mass trespass led to the creation of Britain's first national park in the Peak District in 1951, with the return of at least some of the Duke's land to national ownership.[36]

Many of my Greenham friends went on to engage in other site occupations. The road protests of the 1990s, characterised by permanent encampments, in trees and down tunnels, may not have prevented the building of their immediate targets, such as the M3 extension through Twyford Down. But they challenged and slowed down government road-building programmes and pushed environmental considerations into road planning. Site occupations and obstruction, combined with other forms of campaigning, contributed to the UK government's moratorium on fracking (the process of fracturing shale rock to obtain oil and gas) in 2019. Today, site occupations are one of the most visible components of resistance to HS2, an 'ecocidal, carbon intensive high-speed trainline that will cost the UK taxpayer more than £200 billion', according to the HS2 Rebellion website.[37]

Meanwhile, Greenpeace, over the decades, has built its reputation through nonviolent interventions against environmental crimes, starting with sailing an old fishing boat towards Amchitka Island in Alaska to try and prevent a nuclear weapons test in 1971. The Greenpeace activists quarrelled all the way, got stopped by the US navy before reaching the site and the test went ahead. However, the action aroused massive public interest. Five months later, the Amchitka test programme was cancelled and the island declared a bird sanctuary.

Since then, Greenpeace has grown into a global campaigning organisation with independent national and regional groups based in fifty-five countries. It combines investigative research with lobbying and peaceful direct action, which it defines as:

> Physically acting to stop an immediate wrong at the scene of the crime. Ordinary people around the world can act to confront those in positions of power with their responsibility for stopping global environmental destruction. We act to raise the level and quality of public debate. Above all, we act to provoke action from those with the power and responsibility to make change happen. Guiding all of our actions, always, is a commitment to nonviolence and personal responsibility.[38]

Other effective actions include the Greenpeace ship, *Rainbow Warrior*, trailing Norwegian seal hunters in Scottish waters to prevent the mass killing of seals in 1978. The public outcry resulted in the Norwegians being sent home and seal culling being reduced. Seal culling was completely banned in Scotland in 2020. In the mid-1980s, Greenpeace established a base in Antarctica that monitored the activities of the eighteen Antarctica Treaty signatories, obstructed and prevented the French from building an airstrip, and contributed to an environmental protocol being added to the Antarctica Treaty in

1991. This prohibited mineral exploitation in the region for fifty years.[39]

One of my favourite Greenpeace nonviolent interventions took place recently in 2022, some 200 kilometres off the coast of Cornwall, where there is a marine protected area (MPA), one of a number around the British coast, supposed to protect sea life. However, the problem is that industrial fishing vessels are allowed to drag heavy metal dredges along the sea floor, and this 'bottom trawling' ploughs up the beds, destroys the marine life and releases stored carbon into the atmosphere. Conventional campaigning has failed to stop the activity, so Greenpeace dropped some eighteen large boulders onto the sea bed, which makes it impossible for the bottom trawlers to operate: simple and effective. Similar actions resulted in bottom-towed gear being banned in other marine protected areas, both in Sweden and in the UK. Perhaps they could be banned from all such areas.[40]

Given the achievements of direct nonviolent interventions, I was curious to understand the shifting tactics of different climate protest groups, and particularly Just Stop Oil, whose actions at ten oil terminals in April 2022 actually caused ExxonMobil to briefly suspend operations. However, in the Autumn of 2022 JSO adopted the more publicly disruptive tactics previously used by Insulate Britain. These included two activists suspending themselves from gantries above the Queen Elizabeth II Bridge at the Dartford Crossing, halting traffic for two days. Not surprisingly the hold-ups generated media outrage, as well as the false claim that they had delayed paramedics attending a fatal accident. This was later denied by the South East Coast Ambulance Service.[41]

My GP friend Patrick Hart was involved in the JSO actions at Thurrock oil terminal in Essex. He had joined JSO in fury, on 'the day the UK planning authority announced their decision to overturn local democracy and give the go ahead to Bristol

Airport expansion. 'I was incensed,' he told me. 'Previously I had stuck to a supportive role or low-stakes action hoping this would get the message across. I realised I had to enter into a phase of much higher-stakes action and actively resist the UK government, even if it risked my financial security and liberty.' He felt that JSO were the only group taking serious actions that 'balanced material disruption to the fossil fuel industry' with an 'emotional impact sufficient to generate headlines'.

The aim of the April 2022 action was simple: 'If you won't stop the oil, we will stop it for you. So, day after day, we used different tactics to shut down this oil terminal.' In Patrick's case this involved climbing over the fence and occupying the roof space, sometimes for hours at a time: 'I was arrested five times that week and I was literally sleeping in the cell, getting home, [then] going out on another recce.'

When we talked in February 2023, Patrick had already been found guilty of aggravated trespass, fined and had an injunction imposed on him. He had also had complaints to his workplace and ongoing investigations by the General Medical Council, but he has no regrets and continues to engage in civil resistance. But like many in JSO, he was disappointed at the lack of media impact those actions generated. There were too few people to scale up the actions. So, he understood the change in tactics.

'Shifting tactics is mandatory in this business. You can't just keep doing the same thing again and again. I don't think there's one way here. There are different forms of disruption. One of them is economic disruption, which is what we were trying to achieve by stopping the flow of oil. But another is public disruption.' He thought good actions might incorporate both. 'I don't like marching, I'm not wild on blocking roads,' he said, but 'you've got to have the industry afraid that they're losing money, ... you've got to have the public aware that you mean business.'

Like me, he saw the disconnect between Insulate Britain's actions and home insulation, but pointed out that 'it generated a vast amount of publicity and awareness and discussion'. In November 2022, Chancellor Jeremy Hunt pledged £6 billion to improve energy efficiency and insulate homes in 2025; unfortunately, not the £5 billion a year campaigners said was required, but perhaps a sign that pressure works.

Patrick's involvement had also affected his colleagues. They were beginning to say, 'You've got a point here.' Two years previously a talk he gave on the climate crisis at work generated little interest. But when he started appearing in newspapers and on television he was asked to do another, which had a much more positive response: 'People are like, oh, it must be important then because it's in the media—and they actually give it some thought.'[42]

In April 2022, when Judge Stephen Leake told the twelve Insulate Britain activists he had found guilty that they had inspired him to personally reduce his impact on the planet, Insulate Britain wrote back and said that was not good enough. They called on the criminal justice system to stop criminalising 'peaceful members of the public who are desperately trying to save lives of their families, in their communities, in their country and their global community'. They declared the courts 'a site of nonviolent civil resistance'.[43]

Arguably, juries returning not guilty verdicts in spite of the defendants' admission of law-breaking and judges' clear instructions to find defendants guilty on the basis of the law, have already made them so. These so called 'perverse' verdicts are nothing new and could be regarded as another form of nonviolent intervention. One of the most famous perverse verdicts occurred in 1991, when a jury acquitted Pat Pottle and Michael Randle on charges of assisting the former Soviet spy George Blake to escape from Wormwood Scrubs prison in West London in 1966, even though they both openly admitted they had broken the law in doing so.

The back story to this acquittal deserves some attention. Michael Randle is a personal hero of mine whose own life is a testimony to the many possibilities of nonviolent resistance. Influenced by his father's conscientious objection during the Second World War, he was a conscientious objector to national service. As well as helping organise the first anti-nuclear protest march from London to the AWRE at Aldermaston in 1958, Mike spent a year in Ghana organising protests against French nuclear tests in the Sahara. In 1967, he and others invaded the Greek Embassy in London, to protest the military junta that had seized control in Greece, for which he was sentenced to twelve months in prison.

It was in his role as secretary of the Committee of 100 in 1961 that he was arrested as one of the Wethersfield Six (discussed above). This trial was held in the same year Adolf Eichmann was tried for crimes against humanity for his involvement in the Holocaust. Eichmann's defence was that he obeyed the law; but he was still convicted. Randle, in his defence, argued that since nuclear weapons posed a similar genocidal threat and were also a crime against humanity, he too had a moral obligation to break the law:

> Every individual must finally decide whether millions of lives are threatened by a particular act. There were people in Germany during the Nazi regime who were ordered to commit what have since been defined as crimes against humanity. They would have been going against the law of their country by disobeying their order. I feel they have a moral duty to disobey that order in that situation ... Where I think [the law] is flouting basic human rights I will certainly disobey it, and I feel it would be a moral obligation to disobey ... I feel that the use of nuclear weapons is always contrary to basic human rights. I cannot see any situation in which they would be justified against human beings.[44]

The judge told the jury that motives and beliefs were irrelevant, and directed them to find the six guilty. Even so, it took the jury four hours to reach that verdict. Their recommendation for leniency was ignored by the judge.[45]

It was during his imprisonment in Wormwood Scrubs that Michael met the Soviet spy George Blake. Neither Michael nor Pat had any sympathy for Blake's politics or actions as a Soviet spy. In 1956, Michael had been arrested and deported from Austria for trying to cross the Hungarian border to leaflet and demonstrate his opposition to the Soviet invasion of Hungary. But both felt that the unprecedented forty-two-year sentence imposed on Blake was inhumane and hypocritical. And this was what prompted their next nonviolent intervention. On their release they helped Blake escape from prison by means of a homemade rope ladder, made by Michael's wife, Anne. After accommodating him in various safe houses in London, Michael and Anne then hid him in the bottom of their camper van and drove him to East Germany, where they dropped him off while pretending to go on holiday with their two small boys.

The authorities knew about Pottle and Randle's involvement in this escape in 1970, but chose not to prosecute, perhaps to avoid embarrassment. However, in 1989 they co-wrote and published a book about the whole affair, largely to counter untrue rumours of being communist agents, which were damaging the whole anti-nuclear movement. In 1990 they were arrested and put on trial at the Old Bailey, charged with assisting Blake's escape. They defended themselves. I watched from the gallery as both stood in the dock and persisted in speaking out about their motivations in spite of the judge ruling that the defence of necessity was not available, and frequent admonitions that they should not bring it up.[46]

Michael's final speech to the jury should be recommended reading for all of those doing jury service, and all those considering civil disobedience and risking arrest, which is why I quote it at length. Michael did not dispute the facts, but argued that Blake's 'cruel and unusual punishment', contravened the 1688 Bill of Rights and condemned him to 'slow death':

SORRY FOR THE INCONVENIENCE

A former Lord Chief Justice, Lord Goddard, stated in 1955: 'No one has ever yet been able to find a way of depriving a British Jury of its privilege of returning a perverse verdict'. But 'perverse' needs in many cases to be in inverted commas. The penalty of hanging for stealing sheep was abolished in this country because Juries refused to convict people who were clearly guilty of that offence. As far as the judges and the lawyers were concerned their verdicts were 'perverse'. Today we see that it was the punishment that was cruel and perverse. We are not asking you to act against your conscience. Of course, you must do what you believe is right. But whereas the concern of the lawyers is the law, your concern must be justice.

In 1670, two Quakers, William Penn and William Mead, were accused of causing an unlawful and tumultuous assembly after preaching on a Sunday afternoon in Gracechurch Street in the City of London. The judge directed the Jury to find them guilty and when they refused to do so he locked them up overnight. They were, to quote an account of the time 'without meat, drink, fire or other accommodation: they had not so much as a chamber-pot, though desired'. When they persisted in their refusal to bring in a guilty verdict, he fined each of them 40 marks and sent them to prison. Finally, one of them obtained a writ of habeas corpus and the Court of King's Bench found in their favour. Now I am not suggesting that if you bring in a not guilty verdict his Lordship is going to lock you up. I am saying that, like the Jury in the Penn and Mead case, you have to reach your own independent decision.

An outstanding English judge, Lord Devlin, made this comment on the Jury system: 'The first object of any tyrant in Whitehall would be to make Parliament utterly subservient to his will: and the next to overthrow or diminish trial by Jury, for no tyrant could afford to leave a subject's freedom in the hands of twelve of his countrymen. Trial by Jury is more than one wheel of the constitution: it is the lamp that shows that freedom lives.'

But that is only part of the truth. Institutions alone do not guarantee freedom; that depends crucially on the quality of the people involved in those institutions. The simple existence of the Jury system will not ensure freedom unless Juries maintain the tradition of independence which has made them the scourge of mandarins and bureaucrats down the centuries.

Six years ago, in 1985, a senior civil servant, Clive Ponting, was prosecuted under the Official Secrets Act for disclosing documents to the Labour MP, Mr Tam Dalyell. The documents proved that government ministers had lied to Parliament and the public about the circumstances surrounding the sinking of the Argentinean cruiser *Belgrano* during the Falklands War.

There are important parallels with the present case. First the essential facts were not in dispute; the documents concerned were confidential and Clive Ponting did pass them to Mr Dalyell. Second, the Prosecuting Counsel, Mr Roy Amlot, offered a deal to Mr Ponting before the case opened: if Ponting would resign his civil service post, the prosecution would be dropped. Ponting turned down the offer. He was convinced that what he had done was right, and his public duty.

At his trial he pleaded Not Guilty, arguing that his actions had not been, to quote the words of the Official Secrets Act, 'contrary to the safety and interests of the State'.

The judge, Judge McCowan, in his summing up to the Jury virtually directed them to find Clive Ponting guilty. He told the Jury that the interests of the State were synonymous with the policies of the State and these were determined by the government of the day. Ponting's motives, he said were irrelevant. The Jury should forget about any concept of moral duty.

But the Jury did not forget the concept of moral duty or were unwilling to accept the judge's contention that the interests of the state were synonymous with the policies of the government. They brought in a verdict of Not Guilty.

The lamp of freedom shone more brightly that day, and a dangerous shift towards arbitrary power was avoided. I appeal to you today to keep that lamp burnished and shining and to allow considerations of humanity and common sense to guide your judgment. I invite you to agree with us that what we did was right and to find us Not Guilty.[47]

And they did. Justice prevailed over law. Michael's book about that trial is aptly titled *Rebel Verdict*. Rebel verdicts continue as juries appear to recognise that the government and existing legislation may not be the best protectors of the state, or indeed the planet, and that they have a moral duty to continue to recognise the distinction between law and justice. Arguably, juries are democracy in action at its best. They demonstrate on a small scale what could be achieved by citizens' assemblies. Randomly chosen citizens, given all the available facts and arguments on both sides, and time to consider, are able to come up with a fair judgment.

Some judges are reaching similarly just conclusions. Recently, four young activists who blocked the distribution of oil from Esso Birmingham (an ExxonMobil subsidiary) in April 2022, were found not guilty of aggravated trespass in a Magistrates' Court in Wolverhampton. One of the defendants, Hannah Torrance-Bright, a 21-year-old art student, pointed out the irony of having to justify shutting down ExxonMobil facilities at the same time that the company was on trial in America over allegations that it lied about the climate crisis and covered up the fossil fuel industry's role in worsening environmental devastation.[48]

Troublesome juries and district magistrates that continue to acquit are no doubt one of the reasons that climate activists are increasingly charged with public nuisance, an old common law offence that was made statutory law by the Police, Crime, Sentencing and Courts Act. This requires juries, when considering whether defendants are guilty, to only consider the consequences of the action, not the circumstances in which it occurred. Motivation and beliefs are irrelevant. Judge Silas Reid, the judge who once told the jury that intentionally disruptive action 'need not result in conviction', and they should keep in mind the defendants' 'rights to free speech and to protest', has now moved beyond simply asking juries to ignore these aspects when defendants bring them up. He charges defendants with contempt if they dare to do so.

The 36-year-old care worker David Nixon was the first to be charged with contempt in February 2023. He had tried to bring up insulation and the climate crisis in his closing statement before Reid cleared the court. When given the opportunity to explain and apologise to Judge Silas Reid the following day, Nixon made clear that both the road-blocking action in the City with Insulate Britain in October 2021, and his act of contempt the previous day, were direct nonviolent interventions that he regarded as lifesaving, given the urgent threat of the climate crisis and fuel poverty:

> When I was a care worker sometimes the young people put themselves in a position where they were not safe and me as a care worker would have to intervene and they wouldn't like that and they would tell me to F off. [...] Sometimes I had to intervene to keep them safe and I see very much sitting in the road in similarity to that. And I know I'm annoying people and it's not a nice process but ultimately, I'm trying to keep people safe. I see what I did yesterday in similar terms.

Judge Reid jailed David for eight weeks, saying he found it 'hard to think of a more calculated and deliberate contempt'.[49] He said the same to two other Insulate Britain activists, 65-year-old Giovanna Lewis, and 38-year-old Amy Pritchard, after they too had refused to apologise for ignoring his instructions at their trials and bringing up the deaths caused by fuel poverty and the climate crisis in their closing statements.

In their contempt hearing, Pritchard asked Judge Reid to 'turn your laser-focused attention to bringing justice to the people who are rapidly destroying the conditions we need for life on earth'. He jailed both of them for seven weeks. In both their cases, the jury had failed to reach a verdict after twelve hours of deliberation.[50] Perhaps being prevented by the judge from hearing 'the truth, the whole truth and nothing but the truth' made it impossible to reach a decision.

This was exactly the point Tim Crosland and other human rights barristers and lawyers made in a small demonstration outside Inner London Crown Court on the day of Lewis and Pritchard's contempt hearing. 'What's the point of a jury if we don't trust them with the evidence, if we're going to usurp their role? What does a right to a fair trial mean if you can't speak and explain why you did what you did? What does freedom of expression mean if you're gagged with a threat of prison just for explaining yourself?' Tim asked. 'When Juries and ordinary people and magistrates hear why these actions have happened, they are in general sympathetic. It is not a coincidence that when juries do hear the evidence of climate breakdown and the government's inertia, they're acquitting over and over again.'[51]

When the whole truth is told, it continues to have an effect. Another seven Just Stop Oil protesters appeared in Wolverhampton Magistrates' Court in February 2023, also charged with aggravated trespass for blocking that Esso oil terminal in Birmingham the previous year. Mr Fielding, the prosecutor, initially told them that they were 'self-appointed vigilantes'. But after listening to their testimonies, he retracted the comment and told them they were 'good people'. District Judge Wilkinson allowed all the defendants to explain their motivations and then told them that he had found their evidence deeply moving:

> You are all good people. You are intelligent, articulate and a pleasure to deal with. It's unarguable that man-made global warming is real and we are facing a climate emergency. [...] When the United Nations Secretary-General gives a speech saying that the activity of fossil fuel companies is incompatible with human survival, we should all be very aware of the need for change. Millions of people, and I do not dispute that it may be as many as 1 billion people, will be displaced as a result of climate change. [...] The tragedy is that good

people have felt so much without hope, that you feel you have to come into conflict with the criminal justice system.

He thanked them for opening his eyes to certain things, and apologised for 'sadly' having to convict, ending 'You should feel guilty for nothing. You should feel proud that you care, have concern for the future. I urge you not to break the law again. Good luck to all of you.'[52]

THE WOMAN BEHIND SOLIDARITY: THE STORY OF ANNA WALENTYNOWICZ, FEBRUARY 1984[53]

In November 1982, a woman sat in prison in Eastern Poland writing an open letter to the head of state. In it she asked the now standard questions. Why had Solidarity, their national trade union, been suspended? And why was she, having won the gold cross for her many years of service in the Lenin shipyard in Gdansk, now imprisoned by the same body that had awarded the prize? Was this 'national accord?' She signed it 'Anna Walentynowicz, aged 53 years old, Grudziadz prison.'

The authorities did not reply and kept her imprisoned for an additional four months before bringing her to trial for 'continuing trade union activities in defiance of Martial Law,' a crime punishable by up to 10 years' imprisonment. This time she received a suspended sentence, but imprisonment, or the threat of it, was not new to Anna Walentynowicz.

Endurance, in fact is something she is good at. A widow with one son, she has been struggling with the authorities since she joined the Lenin shipyard as a welder in the Rosa Luxemburg brigade more than 30 years ago. The first time was in 1953 when she complained that women were not getting equal prize money as work incentives. She was arrested and interrogated for eight hours. In 1968 she was fired for complaining about corruption in the government trade unions. It was the start of a long period of harassment and intimidation by the management as she became more and more politically

involved in the 1970s in the development of a free and democratic trade union movement in Poland.

In the critical summer of 1980, she was working as a crane
driver in the Lenin shipyard in Gdansk, respected by her
workmates for her stubborn and defiant attitude. So it was not
surprising that, when she was dismissed yet again while on
sick leave that August, they went on strike to demand her
reinstatement. Other yards struck in sympathy. In two days the
workers at the Lenin yard had achieved their initial demands;
among them: Anna's reinstatement; that of Lech Walesa, who
had also been dismissed; a wage increase; and a promise to
build a monument to honour workers killed in a strike in
December 1970. The strike would have ended there had it
not been for the intervention of Anna and a young nurse,
Alina Pienkowska. Even Lech Walesa, a leader of the strike,
was willing to go back to work. But the demands of the workers at the other factories had not yet been met, and Anna
wanted the strike to continue.

'Alina Pienkowska and I went running back to the hall to
declare a solidarity strike, but the microphones were off,' she
explained. 'The shipyard loudspeakers were announcing that
the strike was over and that everyone had to leave by six p.m.
The gates were open, and people were leaving.

'So Alina and I went running to the main gate. And I began
appealing to them to declare a solidarity strike, because the
only reason the manager had met our demands was that the
other factories were still on strike. I said that if the workers at
these other factories were defeated, *we* wouldn't be safe either.
But somebody challenged me, "On whose authority are you
declaring this a strike? I'm tired and I want to go home." I too
was tired, and I started to cry, like a woman.

'Now Alina is a very tiny person but full of initiative. She
stood up on a barrel and began to appeal to those who were

leaving: "We have to help the others with their strikes, because they have helped us. We have to defend them." Somebody from the crowd said, "She's right." The gate was closed.' That success culminated in the birth of Solidarity in September 1980.

I first met Anna two years later in the midst of martial law. She had just been released from prison. I expected to meet a strong-looking, rather intimidating woman. Instead, the door was opened by a plump, smiling, motherly woman in a blue nylon quilted housecoat. She gave such a cry of delight and welcome that I thought there must be some mistake. However, this is apparently how she responds to all visitors. Her living room was orange-papered and filled with sunlight. There were the furnishings and decorations that I was becoming accustomed to in most Polish apartments: cheap, modern furniture, a divan bed in the corner, rubber plants, a Solidarity poster, and a picture of the Pope. She sat me down with a glass of tea.

She had been in the south of Poland when martial law was first declared in December 1981. 'But I knew my place was at the shipyard, so I came straight back. We were organizing for a national strike, as you know. People were gathering at the shipyard that night, December fourteenth. Barricades were made, and we got the hospital ready. We had no weapons at all. Things seemed to go well from midnight until about five. Then at six o'clock the ZOMO [Polish riot police] started to come through. I tried to walk in front of them. But the workers stopped me. They smuggled me out and I stayed in a private flat.'

She was arrested a few days later. 'There was an old lady in our house, very ill with high blood pressure, I had to take her to the hospital. On the way a car pulled up, some men jumped out and tied my hands behind my back, bundled me into the car and left the old lady standing in the street. By six o'clock that evening I was in Fordon [the women's prison 85 miles north of Warsaw]. By shouting through the doors we discovered that there were thirty-nine of us there, Alina and others.'

Initially, conditions were very bad. 'There were four of us to a two-person cell, with a table on one side of the door and a bucket for a bidet on the other. We were given a limited amount of water once a day at five a.m. to wash, we were allowed ten minutes exercise. There was a lot of food only it was of very poor quality. Of course, if you had money, you could buy better conditions, but we wanted them as a right. So, Anna and four others went on a hunger strike and things got better. They allowed us to attend Mass on New Year's Day, and my son came to see me on the ninth of January. No one had told him where I was.'

Eventually most of the Solidarity women were transferred to Goldapia [Goldap], a summer resort. 'Luxury conditions, but the atmosphere was awful. We had to fight for the right to meet. Internment does have some advantages though; it gives you plenty of time to discuss things.'

I had heard that the government had released all the women internees because they were not seen as a political threat. Anna laughed at this. 'They always said Alina was a better negotiator with the government, because, unlike Lech, she never compromised, and she always got what she wanted.' She wished that there were more women in Solidarity and speculated that their added family demands might have prevented them from being more active.

She saw no need for a separate struggle, however. 'Our main task,' she argued, 'is to create better conditions for everyone, men and women.'

It wasn't until July 1983, after Anna had been released from prison once again, that I managed to see her again and catch up on more recent events. She had been refused entry into the shipyard, but had smuggled herself in. She was arrested after two days, and sent to Rakowiecka prison hospital in Warsaw for psychiatric observation. 'They wanted to prove I was insane. It cost 11,000 zlotys to have eleven specialists to examine me! [The average monthly income in Poland is about 8,000 zlotys.] In fact, it was more interrogation than observation but I told them that I was happy to explain my views and that they should write everything down as I might not get the chance to speak in court. At the end of it all they asked what I thought their opinion might be. I told them it would be totally unjust to describe me as insane. I knew exactly what I was saying and doing. I gained nothing from it, in fact, I was punished, but it was my choice. They agreed.'

Conditions in the prison, however, had not been conducive to sanity. Anna had been put in the male section of the prison and was watched by male wardens. 'The light was left on permanently, the radio blared from seven in the morning to nine at night. You had to stay in your pyjamas and dressing gown the whole time.' Exercise was in a tiny damp yard and the food was thin vegetable soup. 'I wasn't beaten, but others were. One night I heard a boy yelling. So, I banged my hairbrush against the door until someone came. They told me it was an epileptic refusing an injection—but I got a chance to look out and see eight men in uniform beating this man. "I've seen how you treat epileptics" I said. They told me then they'd try to shorten my stay.'

In October Anna was transferred to Grudziadz prison and in March 1983, she was finally brought to trial. 'They held the trial in Grudziadz rather than Warsaw or Gdansk because I had too many friends in those towns.' Even so, supporters came, including Lech Walesa, who testified on her behalf. After a two-day trial she was released with a 15-month suspended sentence.

'Even at the last minute, I thought they wouldn't let me go. They wouldn't open the prison gate. So, I said, "I'm free and if you don't release me this minute I will start screaming." They asked me where my family was. "My only son's in jail" I replied, "but my family is Solidarity, and they are waiting for me." Then the gate opened and there were crowds of people waiting with flowers.'

Now, back in Gdansk in her own flat, she still has problems. 'I can't go back to the shipyard, and everything was stolen from here. I don't get a pension, so I have to work.' She works in a charity office in Gdynia and has written to the shipyard pointing out that she is doing clerical work for which she is untrained for a minimum wage, while they owe her money. She has also involved herself with fighting alcoholism. 'It's easy to rule a nation that drinks. You can do what you want with a drunken man,' she remarked. Meanwhile, her case is still under appeal.

'It's funny,' she mused, 'I started Solidarity, but the winner was Leshek [Lech Walesa's nickname]. He got his job back, but I had to leave after thirty years.' I asked, tentatively, 'Do you think that's right?' 'It's not right, but it's the situation we live in. The men are the public speakers; they have the authority and power. It's part of their make up to feel they are first, and they don't want to share it. A woman is sometimes seen as source of inspiration. But she is more than that, she is a source of strength and she suffers more. A lot depends on her, she has to give time to her home and bringing up children.' 'But you have been working in a man's role,' I argued. 'True,' she said. 'Women have two jobs! Which makes us richer. Men just play with their families, we bring them up and know them.'

'Do you think you'll ever have real equality?' I asked. 'When we have Solidarity,' she replied, smiling. And that, as far as Anna is concerned, is not wishful thinking. 'The new trade unions are worthless. People have tried a little freedom and they aren't going to forget it. These have been two black years, but national consciousness grows in these conditions. People learn to stop being afraid and terror stops being effective. That's why some of the high sentences have been lessened. And people are organising themselves. They employ their friends who can't get jobs. Small enterprises grow.'

She sees education through the church, not violence, as the key. 'When people learn their history, learn what's right, then they start reaching further. Solidarity can't be dissolved. It's a way people look at life.'

We had been talking for two hours, but before I left Anna wanted to show me some pictures. She pulled out an old cardboard box and handed me some battered photographs. Pictures with Anna, a small, smiling figure in a row of sober suited Solidarity men. Anna, radiant, meeting the Pope in Italy. Like some twentieth-century Joan of Arc, I thought. We put the pictures away and walked to the door. 'Work well,' she said, kissing me soundly on both cheeks. 'But then of course you will. Women always do.'

7

WE CAN JUST SAY NO

The strike to reinstate a woman sacked in a shipyard in Poland lit one of the slow fuses that led to the overthrow of communist dictatorships across Eastern Europe in 1989. One woman refusing to give up her seat on a bus in the United States in 1955 helped trigger the civil rights movement of the following decade. One schoolgirl on strike outside the Swedish parliament in 2018 threw open the global discussions and sparked mass movements to address the climate crisis.

Refusal, saying no, non-cooperation, is possibly one of the most powerful nonviolent direct actions available. It can take many forms: strikes, boycotts, divestment, withdrawals. At its heart is the refusal to engage with activities, individuals, businesses or institutions, either to change behaviour or to prevent harm. Again, I like the clarity of Sharp's definition: 'non-cooperation involves the deliberate discontinuance, withholding, or defiance of certain existing relationships—social, economic or political. The action may be spontaneous or planned in advance and it may be legal or illegal.'[1]

In the summer of 1981 I caught a train from Liverpool Street station to Warsaw. I had no idea it was so easy. In *Protest and*

Survive, E. P. Thompson had called on us to reach out directly to colleagues in Eastern Europe and to combine anti-missile protests with a demand for an 'opening of those societies of the East to information, free communication and expression'.[2] I took him at his word and decided to go and visit Solidarity, to witness for myself how mass nonviolent action in the form of strikes had allowed them to create an independent trade union and confront totalitarian rule. The leadership in Gdansk were remarkably friendly to a completely unknown, young British peace activist, although they laughed at the idea of nuclear disarmament. Solidarity spokesman Janusz Onyszkiewicz told me that most Polish people thought the Western peace movement was at best an innocent dupe of the KGB, and that it completely failed to understand the nature of life under a totalitarian regime, from whence came the real threat to peace. We should prioritise fighting totalitarianism. That was what their strikes were doing.[3]

On 12 December 1981, after returning home from a small early demonstration at Greenham marking the anniversary of that NATO decision to deploy theatre nuclear weapons, I turned on the radio at midnight and heard the news. Polish leader General Jaruzelski had declared a 'State of War' against his own people in Poland and declared martial law. In the early hours of the following day, with the help of the Polish army, he brought to an end Poland's attempt at self-limiting revolution. Almost everyone I had met the previous summer was in prison. Perhaps Jaruzelski chose that date deliberately, as if to tell the West their missile threats had played a role in the clampdown.[4]

In the long run, however, martial law and the police state that followed were not successful in bringing the Poles to heel. Resistance was facilitated by the fact that Solidarity was more than just a trade union of 10 million people. As Anna had explained to me, Solidarity had brought Polish society together in a way that enabled them to take care of one another. It had broken down artificial barriers between Poles—'barriers of money, class or caste, of region, of skill and profession [...] and had given back to Poles their feeling of community and common interest'. The main division was now between civil society and the leadership of the Polish United Workers' Party.[5] Communist ideology had lost its last vestiges of credibility in its ruthless crushing of a genuine workers' movement. As a consequence, it was not difficult for society to adopt the strategies recommended by some of the Solidarity leadership, now underground, to avoid a head-on confrontation that might result in massive bloodshed, and, instead to create a strong, informal and decentralised movement that in all sorts of different ways withdrew its support for the Polish state. The idea was to create a parallel society, 'where the government will control the empty shops, but not the market: places of employment, but not people's livelihood: the state media, but not information: printing houses, but not important publications: post office and telephone services, but not communication: the schools, but not education'. This was 'evolution, not revolution'.[6]

In 1984, at the time I interviewed Anna, although everyone was reading underground newspapers, not much appeared to be evolving. But as I continued taking the train to Poland—carrying messages, materials and writing articles—I witnessed the emergence of a new form of non-cooperation. In the winter of 1984, the authorities sentenced a young man, Marek Adamkiewicz, to two and a half years in prison. This was because, when called up for military service, he had refused to take the oath which required him to 'remain faithful to the Government of the Polish People's Republic' and to 'steadfastly guard peace in fraternal alliance with the army of the Soviet Union and other allied armies'. In Poland military training began with school classes at the age of fourteen, and military service was compulsory for all men and some categories of women. All former soldiers remained under military jurisdiction until retirement, and there was no conscientious objector status. Refusal to serve carried a sentence of a possible six years. However, until this point, refusing to take the oath, which had become more common since the imposition of martial law, was usually quietly ignored. Not this time. Perhaps the Polish government thought that punishing a popular, former Solidarity activist would be a warning to others. Instead, more protests followed, including hunger strikes and further refusals. In the autumn of 1985, twenty-five young men sent back their military ID cards with an accompanying letter that stated: 'I am taking this action not because I am opposed to military service, of which the aim is the defence of the motherland, but from a decision not to participate in lies and hypocrisy.'[7]

What the government's repressive tactics had done was foster the emergence of an independent anti-militarist movement: Wolność i Pokój (WiP; Freedom and Peace). The decentralised leaderless network reminded me of Greenham; so did the arguments about whether to focus on the military oath or a wider agenda, which included environmental issues such as challenging the development of a nuclear power plant. But somehow anarchists and punks from Gdansk and devout Christian nationalists from Krakow managed to work out a common programme whose goals included: human rights; national liberation; education on the threat of nuclear war and the need for peace; environmental protection; addressing world hunger and human development; tolerance and promoting nonviolence as a way of life and a tool for struggle.[8] This was a broad agenda, but for many the primary tactic was non-cooperation, as Jacek Czaputowicz, one of the founders of Freedom and Peace, explained in a letter to Western peace movements:

> The majority of us know that it is just as possible to be killed by a truncheon in a police station as by a death-dealing rocket, though death from a weapon of mass destruction is a question of tomorrow,

whereas we face truncheons every day. A man blindly obeying army orders constitutes a greater threat to peace than the neutron bomb. The murderers of Auschwitz and the murderers of Father Popiełuszko excused themselves by saying they were just carrying out orders.[9]

They adopted Otto Schimek as a symbolic patron of the movement. Schimek was a young Viennese Catholic, conscripted into the Wehrmacht and executed in southern Poland in 1944 for 'desertion and cowardice in the face of the enemy'. Local legend claimed that, in fact, he had first sheltered Polish partisans and then refused to participate in their execution. His grave was already a shrine, and anniversaries of his death and birthday were commemorated with a mass before the emergence of WiP. For WiP, 'the significance of this figure extends far beyond the historical limits of the Second World War. The refusal to carry out orders and the following of the dictates of one's own conscience signifies an end to totalitarianism, fascism, communism—to every unacceptable authority.'[10]

Significantly, unlike Solidarity, they chose to sign their own names on documents and to work in the open. They accompanied oath refusal with dramatic tactics such as sit-downs in the centre of Warsaw, throwing leaflets from rooftops and hunger strikes. They demonstrated that there was more political space for action than many in the older generation had realised. They built connections with peace and human rights groups across Europe. Some paid a high price for their actions, receiving long prison sentences. But once again, repression of nonviolent activists had the jiu-jitsu effect. Arrests provoked both national and international protest. In the summer of 1988, the Polish government dropped the reference to the Soviet Union in the oath, and offered some provision for alternative service.[11]

At the same time, Solidarity had begun further rounds of strikes and the Polish government was learning the hard way that it could not solve the now desperate economic situation without the cooperation of society, which meant including Solidarity. Those strikes resulted in the Roundtable discussions that re-legalised Solidarity in early 1989; opened up some press freedoms; and allowed for partially free elections—in which Solidarity dominated and elected the first non-Communist prime minster in Eastern Europe for forty years. Nonviolent anti-communist revolutions followed across Eastern and Central Europe. Non-cooperation had helped to bring the Cold War to an end.[12]

* * *

Non-cooperation is the favoured method of some of the best-known advocates of nonviolence. In a famous essay from 1849 titled 'Civil Disobedience', American philosopher Henry David Thoreau put forward a passionate and eloquent call to refuse cooperation with a government that allowed slavery and persisted in war with Mexico: 'I do not hesitate to say, that those who call themselves abolitionists should at once effectually withdraw their support, both in person and property, from the government of Massachusetts, and not wait till they constitute a majority of one, before they suffer the right to prevail through them.' He condemned voting as a feeble and inadequate expression of wishes: 'Even voting for the right is doing nothing for it. It is only expressing to men feebly your desire that it should prevail.' And he called on others to 'let your life be a counter friction to stop the machine', by refusing to pay taxes and accepting jail, as he was doing, as a means of withdrawing cooperation from a state whose actions he opposed, while accepting the legal consequences:

> I know this well, that if one thousand, if one hundred, if ten men whom I could name—if ten honest men only—aye, if one honest man, in this State of Massachusetts, ceasing to hold slaves, were actually to withdraw from this co-partnership, and be locked up in the county jail therefore, it would be the abolition of slavery in America. For it matters not how small the beginning may seem to

be: what is once well done is done for ever. [...] If a thousand men were not to pay their tax-bills this year, that would not be a violent and bloody measure, as it would be to pay them, and enable the State to commit violence and shed innocent blood. This is, in fact, the definition of a peaceable revolution, if any such is possible.[13]

In a later essay on 'Slavery in Massachusetts', written in 1854, he called for a boycott of those publications that supported slavery: 'The free men of New England have only to refrain from purchasing and reading these sheets, have only to withhold their cents, to kill a score of them at once.'[14]

Mahatma Gandhi, in a speech of 1920, explained in detail what non-cooperation (Satyagraha) meant to him, and the range of activities that it entailed, when he called for this method to be used against the British government:

Is it unconstitutional for me to say to the British Government 'I refuse to serve you'? Is it unconstitutional for our worthy Chairman to return with every respect all the titles that he has ever held from the Government? Is it unconstitutional for any parent to withdraw his children from a Government or aided school? Is it unconstitutional for a lawyer to say 'I shall no longer support the arm of the law so long as that arm of law is used not to raise me but to debase me'? Is it unconstitutional for a civil servant or for a judge to say, 'I refuse to serve a Government which does not wish to respect the wishes of the whole people'? I ask, is it unconstitutional for a policeman or for a soldier to tender his resignation? [...] I am here to plead [...] for real substantial non-cooperation which would paralyse the mightiest Government on earth.[15]

Martin Luther King explained in a speech in 1961 why non-cooperation was actually a foundational aspect of civil disobedience. He justified the students' actions of the previous years in the civil rights movement, including boycotts and freedom rides, arguing that 'it is as much a moral obligation to refuse to cooperate with evil as it is to cooperate with good'. It was this principle which underlay the responsibility to obey just laws such as the 1954 Supreme Court's decision that segregation in schools was unconstitutional; 'and at the same time, we would disobey certain laws that exist on the statutes of the South today'. Because 'there are two types of laws. There are just laws and there are unjust laws. [...] any law that degrades the human personality is an unjust law.' Civil disobedience, for King, was peacefully refusing cooperation with immoral and unjust laws and accepting the penalties for doing so.[16]

In fact, the long history of powerful examples of non-cooperation campaigns predates these writers. They include simple and legal actions, such as the sugar boycott organised by the Methodist schoolteacher and abolitionist Elizabeth Heyrick. In 1824 she wrote a pamphlet arguing that everyone was complicit in slavery:

> The perpetuation of slavery in our West India colonies is not an abstract question, to be settled between the government and the planters; it is one in which we are all implicated, we are all guilty of supporting and perpetuating slavery. The West Indian planter and the people of this country stand in the same moral relation to each other as the thief and receiver of stolen goods.

She also argued (in language that might be familiar to activists campaigning against fossil fuels today) that the gradualist parliamentary methods of the official anti-slavery society were too slow and ineffectual, and that it was 'high time [...] to resort to other measures, to ways and means more summary and effectual'. These included an end to taxpayer subsidies to West Indian plantation owners and an immediate boycott of sugar: 'That abstinence from West Indian sugar alone, would sign the death warrant of West Indian slavery, is morally certain.'[17]

William Wilberforce and other leaders of the official Society for the Mitigation and Gradual Abolition of Slavery found the idea of female activists difficult to countenance, and tried to suppress the pamphlet. But female anti-slavery societies began to spring up across the country, meeting and distributing the pamphlet and promoting the boycott. In 1830, Heyrick once again demonstrated the effectiveness of a threatened boycott, when, as treasurer of the Female Society in Birmingham, she suggested they withhold their annual £50 donation to the national Anti-Slavery Society until 'they are willing to give up the word "gradual" in their title'. It worked. The term 'gradual abolition' was dropped at the national conference the following month and the society agreed to campaign for immediate abolition.[18]

The word 'boycott' actually comes from another non-cooperation campaign, again with women in a leading role. Farmers in Ireland refused to pay rent, harvest land or take land from which others had been evicted as part of the National Land League's struggle against the British. In 1880, the Ladies' Land League in County Mayo initiated a programme of social ostracism against Captain Charles Boycott, the local English land agent. Shops would not serve him, people would not speak to him, and normal services were unavailable, forcing him to leave for England. The idea and the name caught on and spread through Ireland. In 1881 the British government passed a comprehensive Land Reform Bill.[19]

In occupied Norway during the Second World War, a social boycott of German soldiers, including staring straight through them and refusing to sit beside them on public transport, was so effective that the fascist authorities made it an offence to stand on a bus if there was an empty seat.[20] It was Norwegian schoolteachers in occupied Norway who carried out one of the most powerful and inspirational non-cooperation campaigns. When the Nazi-installed fascist 'minister president' Vidkun Quisling, as part of his attempt to create a Corporative State, tried in early 1942 to create a National Socialist teachers' union and introduce Nazi ideology into schools, the majority of teachers refused to join the union or teach the official curriculum. The closure of schools, withholding of salaries and imprisonment of 1,300 teachers, some in a concentration camp in the Arctic north, had no effect. Norwegian society rallied round the teachers, with children taught privately and salaries paid by the Resistance. In November 1942, Quisling gave in. The teachers were released and schools reopened. This successful nonviolent resistance was the first step in bringing down Quisling's collaborationist government.[21]

Boycotts were also used with great effect throughout the civil rights campaigns of the 1950s and 1960s in the Southern United States, not just in the well-known example of Montgomery, but as complementary actions to other nonviolent interventions. For example, in Nashville in 1960, Black citizens decided not to buy any new Easter clothes, in support of students doing lunch counter sit-ins. A month later, six downtown stores integrated their lunch counters.[22] Boycotts of downtown businesses were also a significant part of the Birmingham, Alabama campaign in 1963, as I have described in Chapter 2.[23] And in 1964 and 1965, prefiguring the divestment movements of later decades, the National Association for the Advancement of Colored People (NAACP) wrote to American brokerage firms, urging them not to buy, sell or invest in bonds from Southern states that abused civil rights, as a result of which some firms did divest.[24]

In the same year, the National Farm Workers Association (NFWA) in California began a five-year nonviolent struggle, led by César Chávez, for farm workers to have the same basic rights as other occupations, including the right to form unions. They combined strikes and a powerful long-distance march with a nationwide boycott of Californian grapes. These actions led to union contracts being introduced in 1970.[25]

The fact that we know what is going on inside parliament is thanks to courageous acts of direct intervention and non-cooperation with an unjust law. Both houses of parliament banned the reporting and publication of parliamentary debates in the seventeenth and eighteenth centuries. However, in 1768, a certain John Almon began publishing them, and other papers followed suit. When the printers were arrested and brought before magistrates, two refused to convict and were imprisoned in the Tower of London. There was an outcry, with street demonstrations against both prime minister and King. The magistrates were released and parliamentary proceedings continued to be published without penalty from 1771.[26]

'Perverse' verdicts, as described in the previous chapter, could also be seen as one form of non-cooperation by juries. Another method juries used was to undervalue stolen goods by a penny or a shilling lower than the value that would have resulted in capital punishment; or refuse outright to convict for crimes against property. In 1861, capital punishment was abolished for property crimes.[27]

Given the power of these actions, it is interesting how focused both activists and the media are on disruptive street protests today. I agree with Michael Randle, who argues that 'it is in fact extensive non-cooperation which is more likely to render a policy impossible to implement'. Moreover, as he points out:

> Refusal on principle to cooperate is probably the least contentious form of civil resistance, the one most readily accepted as justified within democratic systems. The safeguard here for the democratic process is that non-cooperation does not begin to be coercive unless it is taken up by very large numbers of people.[28]

And if large numbers do engage in non-cooperation within a democracy, that does reflect a body of opinion which needs to be addressed. A good example in this country is the refusal of large numbers of British citizens to pay the poll tax introduced by Margaret Thatcher in 1990, because they considered it unjust. Arguably, it was not the dramatic riots of March 1990, but this non-cooperation by an estimated one in five rate payers, that led to the withdrawal of the tax, and contributed to Thatcher's resignation that November.[29]

In mid-January 2023, I was standing on a picket line outside our local urgent care centre, supporting nurses and healthcare assistants on an official strike. The nurses standing beside me held banners saying: 'People aren't dying because nurses are striking; Nurses are striking because people are dying.' The patients going in and out of the unit gave thumbs-up signs and many of the cars driving by honked in support. The nurses were maintaining night-shift levels of care, and in any emergency, more would respond. Sue told me she qualified five years ago, earned £25,000 a year and was £42,000 in debt, because nurses' bursaries were stopped just when she entered nursing. The healthcare assistant beside her earned £12 an hour. 'I can earn as much working at Sainsbury's or MacDonald's, so I do wonder why I stay', she told me.

A teaching assistant at the further education college up the road had made the same point when I had gone to support their picket line the previous December. Julie, a learning support assistant, told me she was paid £9.70 an hour for helping children with the most complex difficulties. That did not include the almost two days a week of extra unpaid hours that many worked as a matter of course. The previous day, Julie had been helping a child who was brain-damaged after an accident a few years back. Simply moving her large wheelchair from class to class was a major undertaking. She also needed speech and language assistance and physiotherapy, all of which Julie provided. Another of her children had severe autism and was really upset that she was on strike. In fact, Julie was leaving the job because it was unsustainable for her. There were colleagues depending on food banks, and one of her friends was foraging in the countryside because she had run out of food. As with the nurses, the public walking by and the drivers passing were supportive, most honking their horns and waving.

Meanwhile the boss of BP doubled his earnings to a record £10 million while his company made a £23 billion profit in 2022, thanks to the war in Ukraine. Global Witness reported that the five largest private-sector oil and gas companies: Chevron, ExxonMobil, Shell, BP and TotalEnergies made profits that amounted to US$195 billion in 2022, 'nearly 120% more than the previous year, and the highest level in the industry's history'. ExxonMobil came top, reporting an annual net income of US$56 billion.[30]

Again, I am filled with a kind of wondering fury at how it has come to this. Why do we continue to allow the companies producing those greenhouse gas pollutants that are known to be killing us by the day, both directly and indirectly, to make such obscene profits, and reward their bosses so highly, even as they inform us that they plan to increase their oil and gas production in the next few years, and not to reduce it? And why at the same time do we give so little value to the people who look after us at our most vulnerable and frailest? The government's initial answer to this discrepancy has been not to tax the profiteers and pay the staff who run our public services a decent wage, but to introduce anti-strike legislation that limits their right to strike.

Health workers and teachers were not the only ones taking industrial action. Strike action, or the threat of it, in 2022 and 2023—by lawyers, bus drivers, telecommunications workers and Scottish health workers—won higher pay.[31] So strikes were having some effect, and many of my friends and colleagues wanted them to continue, especially as deals offered by the UK government actually amounted to pay cuts in real terms. Would oil executives find that acceptable? Others asked why we didn't go further, beyond wage increases, to demanding a reform of the system itself?

But how do you go on strike against the climate crisis? How do you strike to protect nature? The school strikes drew attention to the issues, and in April 2022 I witnessed how the courageous action of one individual can push the dial further, when I met Angus Rose, sitting outside parliament wrapped in a blanket. He was already in the fourth week of a hunger strike. His demand was simple; he wanted the cabinet to have a televised briefing on the climate crisis by the government's Chief Scientific Adviser, Sir Patrick Vallance. He had been inspired by the Swiss hunger striker Guillermo Fernandez, whose thirty-nine-day hunger strike resulted in Swiss lawmakers agreeing to have a briefing from environmental scientists.[32] Angus, a software engineer, told me that he was acting on behalf of his nephews and niece, and that he was prepared to die because the situation was desperate. 'If I was in Ukraine, I'd be risking my life to protect my nephews and niece, but when something that's orders of magnitude more significant and people don't understand that, we have to mobilise, right? It's so urgent, yet even when people understand it, they just don't want to look up', he said, referring to the movie.[33]

What was impressive was how this singular action by one person galvanised others. Angus already had some forty people supporting him, checking on his wellbeing, making media contacts and joining in fasting. Seventy-nine scientists had written an open letter supporting the demand, as did twenty-nine eminent health professionals.[34] And there was a result. Angus ended his strike on day thirty-seven, when Green MP Caroline Lucas arranged for Sir Patrick Vallance to brief the All-Party Parliamentary Group on Climate Change. Two briefings followed, although sadly only a minority of MPs attended.[35]

Can we do more? In 2014, Archbishop Desmond Tutu, revered veteran of the anti-Apartheid movement in South Africa, pointed to the dangers of climate change and asked: 'Who can stop it? Well, we can, you and I.' He reminded us that it was the economic and moral impact of boycotts, divestment and sanctions that had played a major role in ending Apartheid, and called for a similar campaign to address the climate crisis:

People of conscience need to break their ties with corporations financing the injustice of climate change. We can, for instance, boycott events, sports teams and media programming sponsored by fossil-fuel energy companies. We can demand that the advertisements of energy companies carry health warnings. We can encourage more of our universities and municipalities and cultural institutions to cut their ties to the fossil-fuel industry.[36]

One of my earliest memories of political action was my mother refusing to accept a bottle of South African wine that a guest had brought to a dinner party. Later she explained to me what Apartheid was, and how boycotting Cape-grown oranges and anything else made in South Africa could make a difference. The international boycott campaign which included consumer, sporting, cultural and academic boycotts ran for thirty-five years and was one of the most powerful components of the British anti-Apartheid movement. Anyone could do it and hundreds of thousands did. In June 1986, an opinion poll found that 27 per cent of people in Britain boycotted South African products.[37]

The simplicity of an individual boycott and its potential knock-on effects was beautifully demonstrated by Mary Manning, a 22-year-old cashier at Dunnes Stores on Henry Street in Dublin in 1984. She followed her union's direction not to handle South African goods and refused to cash two South African grapefruits. She was supported by her colleague, union shop steward Karen Gearon. When the shop management could not persuade them to back down, they were suspended, whereupon twelve other workers went on strike. The strike lasted almost three years, during which time the workers spoke at the UN, and Desmond Tutu invited them to South Africa, where they were refused entry and deported, gaining international attention. In 1987 the Irish government introduced a ban on all South African products. Nelson Mandela was released three years later, and personally thanked the strikers when visiting Dublin that year.[38]

We can all do something. The first political speech I ever made was to my student union, asking them to back a motion calling on the university to divest from any companies involved in South Africa. It passed. My first pay cheque as a trainee GP in Cornwall was from Barclays Bank, which had a significant subsidiary in South Africa. I asked my colleagues if we could move the practice's account. A meeting was held, they agreed and wrote to Barclays explaining why. That small action clearly counted, given the incensed response we got from the bank. We were just one small part of the Boycott Barclays campaign. Students refused to sign up to new accounts; universities, charities and local authorities closed theirs. In 1985, a shadow report claimed that since 1980, Barclays had lost accounts with an estimated annual turnover rate of more than £6,000 million. In November 1986, Barclays sold its stake in Barclays National, its South African subsidiary.[39]

I told this small story at our town council meeting in March 2023. Jan Power, my councillor friend, had put forward a motion that the council should, along with other divestment actions, move its bank account from Barclays because they were the British bank with the largest investments in fossil fuels. There had been a few recalcitrant voices in the finance committee. Jan thought a show of support from the public would be helpful. So a group of us went along and were allowed our say at the beginning of the meeting. The motion to move the account passed unanimously.

Out on the street it's a similar story. My Dolphins affinity group friends and I go out on alternate Saturdays and play 'How Green is your Bank?' on the main shopping street. We have a large board that ranks British banks on a coloured bar chart according to how much they are investing in fossil fuels. We cover up the names of the banks and ask players to place a magnet with their bank's name on the bar that indicates how much they invest. Barclays, Santander, HSBC, NatWest and Lloyds provided £12.9 billion to the fifty largest oil and gas companies, all of whom are still expanding production; so dark red to orange-coloured bars for these worst offenders, light to dark green for the best. I'm surprised at how easy it is to engage people in the game, and at their interest in the discussion that follows. Perhaps it's the chocolate biscuit prizes (everyone wins a prize); the new, sometimes surprising information that £1,000 in a current account with Barclays contributes more than a quarter of a ton of carbon to the atmosphere in a year; and the simplicity of the action that can be taken to address it. No need to glue yourself to the road.[40]

Individual banking boycotts are just one small part of the wider divestment campaign. The Royal College of Psychiatrists added fossil fuels to its 'excluded investments', which include tobacco, arms, adult entertainment, alcohol and gambling, stating that it would now only 'invest in companies that follow good environmental, social and governance policies and practices', although at the time of writing the College has yet to move its bank account from Barclays and is being lobbied to do so. To date, six of the royal medical colleges have committed to full divestment, as have 101 British universities, both part of a global institutional divestment movement that according to the Global Commitments Divestment database will remove US$40 trillion from the fossil fuel industry.[41]

Does it have an effect? Some argue that the divested capital is simply sold to others, so nothing is achieved. But this misses the point that a divestment movement does more than remove money. It can be invested in companies with ethical sustainable goals, and the action educates, stigmatises and creates financial uncertainty around fossil fuels, meaning that, as David Carlin wrote in Forbes in 2021:

> Fossil fuel companies are valued on their reserves. Climate science
> tells us that to maintain a safe climate, most of those reserves must

remain in the ground. Unexploitable reserves become worthless, stranded assets. [Thus] Financial institutions can choose divestment to avoid major losses and gain the opportunity to reinvest in more promising industries.[42]

In 2019, just before he stepped down as governor of the Bank of England, Mark Carney warned that 'Companies that don't adapt—including companies in the financial system—will go bankrupt without question.'[43] According to a report by Insure Our Future, more and more insurance companies are refusing to underwrite new coal plants and mines, as well as oil and gas expansion, suggesting that they too recognise that fossil fuel expansion is financially risky.[44]

Public pressure on companies involved in supporting fossil fuel projects works. The 1,443-kilometre East African Crude Oil Pipeline (EACOP) from Uganda to Tanzania, if built, would displace more than 100,000 people, severely damage water resources, wetlands and biodiversity hotspots, while generating 34 million tons of carbon emissions each year. It is already damaging communities in its projected path, while frontline activists have been intimidated and arrested. In October 2022, nine Ugandan students were detained and charged with public nuisance, simply for marching to the European Union offices in Kampala in support of the European Parliament's resolution condemning the pipeline.[45] By November 2022 twenty-four banks and eighteen insurance companies had pulled out of the project.[46]

* * *

Diary: 25 February 2023, London

I am standing in the main entrance hall of the British Museum in front of the rotunda holding one section of a banner. Mine has a large weeping eye on it, my neighbour's banner says 'DROP BP'. There are other sections showing clasped hands, an oil rig, a flower. We are just a small group. The action is part of the continuing campaign of creative protests organised within the museum by Culture Unstained to get the British Museum to drop BP as a corporate sponsor. We take it in turns to read out our letter to the director:

> *Despite reporting record-breaking profits of £23 billion just weeks ago, BP has also slashed its target to reduce fossil fuel production by 40% by 2030 to just 25% by 2030; and reduced its 'low carbon spend' from 2021–22 while earmarking up to £6.2bn for oil and gas projects. [...] The British Museum is an institution founded on a dark legacy of colonialism and empire; enabling fossil fuel titans to continually pursue extraction and destruction in the Global South simply extends the dark shadow of imperialism and white supremacy in new forms. [...] As long as BP is welcome inside the British Museum. [...] we will not be silent. We stand in solidarity with frontline communities, as well as the archaeologists, historians, cultural figures, and your own striking workers who want more from the British Museum.*

I am astonished that we have been able to go ahead. There was an hour-long queue stretching down the street and round the corner just to get in to the museum, because the security searches of bags have become so thorough. They appear to be looking for cans of soup, and glue. Then, when we gathered by the BP-sponsored BP Lecture Theatre, we were informed by security guards that this was not a public area. I think our plain black clothes were a giveaway. When we all moved to the obviously public restaurant area, they searched us all over again. But after we walked out in front of the rotunda, facing the entrance, they left us alone as Phil pulled bits of banner from under his shirt and gave each of us a piece. And once we were in place, some of the security guards smiled and gave a thumbs up, taking photos on their mobiles, perhaps because the action had been deferred in support of their strike the previous week. The public seem to like us, stopping to listen to the letter being read aloud and taking pictures. Some join us as we process around the main hall. The only hostility we meet is from the guard who will not let us deliver our letter to the museum director, and responds to our request that he hand it in for us with 'No I'm not touching it, definitely not.'[47]

* * *

The sponsorship by BP and other fossil fuel companies of art and culture, sporting events and funding of academic research in universities, allows them to pretend to be on the side of the good guys while continuing business as usual. It's rather like the betting companies running Gamble Aware advertisements. This is why pressuring institutions to drop their sponsorship and funding is as important as divesting. The Science Museum's continuing relationships with fossil fuel companies including BP and Equinor and the coal-producing conglomerate Adani, have become increasingly controversial. In April 2022, the museum had to cancel a panel discussion at the last moment because two of the speakers withdrew when they learnt of the fossil fuel sponsorship. In the last two years, prominent scientists have refused to work with the institution until it cuts its ties with these companies. Two researchers in global plastic pollution would not allow their work to be exhibited in the permanent collection, while other eminent scientists have resigned from the museum's boards or refused to become trustees. Mathematician Dr Hannah Fry explained her reasons for resigning from the board in an article in *The Times*:

> By allowing such public ties with these companies, I worry that the Science Museum gives the false impression that scientists believe the current efforts of fossil fuel companies are sufficient to avoid disaster. [...] In the last week, the museum has reacted defiantly amid the reasonable voices calling for change [...] This is a debate where young people are leading the charge, and I cannot in good conscience remain in post while the museum is not proactively engaging with the very people it was built to inspire.[48]

In fact, the Science Museum and the British Museum were both holdouts against a growing tide of refusals of such sponsorship.

In the UK this included Tate, the National Gallery, the National Theatre, the British Film Institute and the Royal Shakespeare Company (RSC). In January 2023, the Royal Opera House ended thirty-three years of BP sponsorship.[49]

All these withdrawals illustrate the importance of pressure from all directions: the resignations of public figures, creative protests, letter writing. In June 2019 the actor Mark Rylance wrote an open resignation letter to the RSC asking 'BP or not BP? That is the question', and explaining that he was resigning because of the company's continuing acceptance of BP sponsorship. Their argument that without BP the £5 ticket deal which they sponsored for young people would be cut, was answered by Culture Unstained crowdfunding alternative £5 tickets. Rylance pointed out that BP had, like ExxonMobil, known about the dangers of fossil fuels for thirty years, yet lobbied against climate action, and that 'if oil companies were taxed properly this could not only pay for greater public funding of the arts, sciences and health services, but could support the rapid transition to green energy we desperately need'. He felt he had to resign because 'I do not wish to be associated with BP any more than I would with an arms dealer, a tobacco salesman or any company or individual who wilfully destroys the lives of others, alive and unborn. Nor do I believe would William Shakespeare.'[50]

That September, youth climate strikers wrote a powerful letter to the RSC threatening their own boycott:

> We are the audiences of the future and we will not support theatre that accepts sponsorship from a company that is continuing to extract fossil fuels whilst our earth burns. [...] You can be sure that if you continue to accept funding from BP we will come to Stratford-Upon-Avon, not to see a show, but to make a scene ourselves.

They pledged not to go to RSC plays, and to lobby their families, friends and schools to stay away until the company dropped BP.[51] A week later the RSC announced that BP sponsorship would conclude at the end of 2019, two years earlier than planned, and stated that 'Amidst the climate emergency, which we recognise, young people are now saying clearly to us that the BP sponsorship is putting a barrier between them and their wish to engage with the RSC. We cannot ignore that message.'[52]

Academics are also calling for boycotts. In March 2022, some 500 researchers wrote an open letter calling for UK and US universities to reject research funding from fossil fuel companies because such funding 'represents an inherent conflict of interest, is antithetical to universities' core academic and social values, and supports industry greenwashing. Thus, it compromises universities' basic institutional integrity, academic freedom, and their ability to address the climate emergency.' They pointed out the parallel with numerous institutions rejecting money from the tobacco industry to fund public health research, because of that industry's record of spreading disinformation: 'Today, the fossil fuel industry has employed disinformation tactics from the same playbook, working to sow doubt about climate science, silence industry critics, and stall climate action.'

By March 2023, some 800 academics had signed the letter.[53] Princeton University became one of the first universities to actively both divest and dissociate from 'fossil fuel companies that participate in climate disinformation campaigns or otherwise spread climate disinformation, and from companies in the thermal coal and tar sands segments of the fossil fuel industry unless they prove able to meet a rigorous standard for their greenhouse gas emissions'.[54] Meanwhile Birkbeck, University of London, has stopped fossil fuel companies recruiting through its career service, and hopes others will follow its example.[55]

One kind of action leads to another. I watch again the widely circulated video on YouTube, where Caroline Dennett explains why she is resigning as a safety consultant from Shell, after working to help them with individual safety for eleven years. She was already engaged in climate activism and giving talks 'asking councils to take their money out of fossil fuels but then admitting I work in the industry and need to do this myself. We're all hypocrites to some extent but this was starting to feel ridiculous.' Then some Extinction Rebellion activists glued themselves to the furniture inside the Shell headquarters, and their sign, 'Insiders Wanted', caught her eye. A few weeks later she posted her letter and video on LinkedIn:

> I can no longer work for a company that ignores all the alarms and dismisses the risks of climate change and ecological collapse. Because, contrary to Shell's public expressions around Net Zero, they are not winding down on oil and gas, but planning to explore and extract much more. I want Shell execs and management to look in the mirror and ask themselves if they really believe their vision for more oil and gas extraction secures a safe future for humanity. We must end all new extraction projects immediately and rapidly transition away from fossil fuels, and towards clean renewable energy sources. Shell should be using all its capital, technical and human power to lead this transition, but they have no plan to do this.

She ends by asking colleagues: 'Join me, and exit the industry if you can.'[56]

We all have to find an exit. We are in a mess. The IPCC synthesis report pulls together the key findings of the global community of scientists over the last six years and it is unequivocal. 'Human activities, principally through emissions of greenhouse gases, have unequivocally caused global warming.' The damage is being done right now, to human beings, to ecosystems on land, and it is the most vulnerable people, some 3.5 billion of them, who have contributed the least to the problem, who are suffering the most.[57]

Melinda Janki spelt it out for me when we met, pointing out that the carbon bomb that ExxonMobil was exploding in Guyana was not of Guyana's making: 'Guyana is a carbon sink. ExxonMobil is an American problem that's landed in our laps. You cut your fossil fuel use in the United States and the problem disappears. The US is an annual carbon bomb.' It is our Western lifestyle that is ecocidal:

> That's what has to be stopped. And there's no easy way to do this. I mean, I have lots of campaigning friends in the north who really don't make a connection between their carbon footprint and what's happening in the rest of the world. They all talk about climate change as if it's out there, but they don't think about the fact that they drive, they use all these machines in their houses because they can't manage without them. Manage without them! Wash your dishes, instead of switching on the dishwasher![58]

Some of my more radical friends argue that it's not about individual change; if you don't engage in collective civil resistance to change government policy, then you are part of the problem because it is only systemic change enacted by governments confronting corporate power that can change this situation. My own view is that in this power to say no, to refuse cooperation with the companies, the institutions and systems doing us harm, there is space for everyone, whoever and wherever they are, to act, and for it to have a collective effect. A new initiative called 'Take the Jump' makes a similar argument, stating that 'it is not up to citizens to "save the world" on their own, government and business still have the largest responsibility, for up to 73–75% of emissions.' However, 'citizens have primary influence over 25–27% of the emissions savings needed by 2030 to avoid ecological meltdown. This is the first time this impact has been quantified and shows citizens are not powerless.'[59]

'Take the Jump' advocates individuals choosing and trying out what are basically simple acts of boycott and non-cooperation, all of which reduce consumption: not flying more than once in three years, eating less meat, keeping electronic goods for at least seven years, not buying more than three items of clothing per year, moving your money to more ethical accounts. As I discovered out on the street in Penzance, it's a way to begin. And if you ask your neighbour, friend, community, workplace to do the same thing, you start a movement. I hope Melinda would approve.

Non-cooperation is everywhere. In Aylesbury, 72-year-old Dr Jane McCarthy withholds her council tax as part of a campaign to get the council to divest its pension fund from Barclays. Summoned to court for non-payment, she explained: 'as a Quaker and a grandmother I cannot hand over money that is being invested in the destruction of our beautiful planet, creating incredible harm and climate injustice right now, and appalling consequences for all future generations'.[60]

And Gary Lineker's friends dramatically illustrated the power of a social boycott when the BBC suspended its star football commentator for tweeting that Suella Braverman's Illegal Migration Bill was 'an immeasurably cruel policy directed at the most vulnerable people in language that is not dissimilar to that used by Germany in the 30s'. Colleagues, fellow pundits, commentators and show hosts lined up to say they would not take part in that day's football programming. The result: a weekend of football shows was cancelled or adapted. In the face of this boycott, the BBC backed down and Lineker was put back on air, while the BBC reviewed its guidelines.[61]

Meanwhile, Tim Crosland at Plan B, along with a number of other colleagues including Melinda Janki, and with support from the Good Law Project, formed a group called 'Lawyers Are Responsible' in December 2022. They published a 'Declaration of Conscience' at the end of March 2023. The Declaration makes clear that the signatories 'believe in upholding the rule of law, as a cornerstone of social stability, prosperity and democratic values'. They note the UK government's continuing approval of a coal mine and granting of new licences for oil and gas production, in spite of having declared a climate emergency and acknowledged that we face an existential threat, even though

The International Energy Agency concluded that there could be no new oil or gas fields or coal mines if the world was to reach net zero by 2050 [...] The likely consequences of 'overshooting' 1.5C include widespread loss of life and livelihoods, catastrophic harm to health, large scale population displacements and the destruction of critical infrastructure [...] Climate change results in climate injustice: lower-income groups and countries and the younger generation have contributed least to climate change but are the most severely affected by it.

The Declaration expresses 'grave concern that the above developments pose a serious risk to the rule of law', and notes that 'lawyers who support transactions the effects of which are inconsistent with the 1.5C limit contribute towards the above consequences'. It then goes on to call for, among other things, 'the implementation of the polluter pays principle to correct the market failure that externalises to society the costs of greenhouse emissions created by burning fossil fuels'. It supports 'individuals' democratic right of peaceful protest, in particular peaceful protest aimed at drawing public attention to the climate crisis', and notes concern about new legislation that restricts those rights. Most significantly it ends with a call for non-cooperation: 'We declare, in accordance with our consciences, that we will withhold our services in respect of: (i) supporting new fossil fuel projects; and (ii) action against climate protesters exercising their democratic right of peaceful protest.'[62]

The launch was marked by a short film with some of the signatories projected onto the walls of the Royal Courts of Justice. I watched Tim Crosland speaking on Twitter: 'Behind every new oil and gas deal sits a lawyer, and behind that lawyer sits a law firm getting rich on those profits while other people die.'[63]

I was reminded of Thoreau pointing out, nearly 200 years ago, that the greatest obstacle to ending slavery was not politics in the South but commercial interests in the North of the US: 'A hundred thousand merchants and farmers here, who are more interested in commerce and agriculture than they are in humanity, and are not prepared to do justice to the slave and to Mexico, cost what it may.'[64] The same could be said today; just substitute 'lawyers and fossil fuel executives' for 'merchants and farmers', and 'life on this planet' for 'slave' and 'Mexico'.

There was the predictable twitter storm. *The Daily Mail* was very upset, giving a whole front page to an attack on 'woke lawyers' who break 'the cab rank rule'.[65] This is part of a barrister's professional code of conduct, which obliges them to take on any case for which they have received instructions and have competence, regardless of the identity of the litigant, the nature of the case and the personal beliefs of the lawyer, because everyone is entitled to legal representation. Tim told me when we met that the rule was a 'wonderful fig leaf' hiding the British legal profession's entanglement with the carbon economy. He hoped the Declaration would 'tear the fig leaf away'.[66]

Moreover, as another signatory, Jolyon Maugham KC, director of the Good Law Project, wrote in *The Guardian*, although the rule is a 'beautiful idea' it 'is often used, by those whose professional lives demonstrate no interest in access to justice, to shield barristers from criticism for self-interested decisions to act for wealthy rogues. In the real world it is money, not character, that divides those who do and don't get to use the law.' He ended with a powerful call for non-cooperation with bad laws:

> Sometimes the law is wrong. What it stands for is the opposite of justice. Today's history books speak with horror about what the law of yesterday did, of how it permitted racism, rape and murder. And tomorrow's history books will say the same about the law as it stands today, of how it enabled the destruction of our planet and the displacement of billions of people. [...] We should not be forced to work for the law's wrongful ends by helping deliver new fossil fuel projects. We should not be forced to prosecute our brave friends whose conduct, protesting against the destruction of the planet, the law wrongly criminalises. That is a beautiful idea, too.[67]

DIARY

NOVEMBER 1985, GUATEMALA CITY

'Could you start escorting tomorrow?' Alice asks me. 'We are in a bit of a crisis, two of our volunteers have had no break for twenty-four hours.' Alice is from the Canadian office of Peace Brigades International (PBI) and is on a list of eight volunteers who are supposed to be deported, because they are accused of 'influencing' GAM. My stomach tightens. This was not what I expected. I thought I was coming to Guatemala for a few weeks to write an article about the Peace Brigades and their work with the Grupo de Apoyo Mutuo (GAM: Mutual Support Group) before I joined an international peace march across Central America. I definitely wanted to help.

I heard about GAM and what was happening in Guatemala from someone passing through Greenham. Corrupt military dictatorships of one form or another have been waging war on their people since 1954, when a small landholding oligarchy and the armed forces, with the support of the CIA acting on behalf of the USA-based United Fruit Company, overthrew the government of Jacobo Arbenz which was delivering land reform. The violence has taken the form of assassinations and forced disappearances, and as a guerrilla movement emerged in the countryside, brutal counter-insurgency operations in which whole villages are razed to the ground in the name of countering 'communist' subversion. In 1984 alone there were on average 100 political assassinations and more than 40 disappearances a month, according to the British Parliamentary Human Rights Group who visited last year.

SORRY FOR THE INCONVENIENCE

The GAM was founded by Nineth Montenegro de Garcia at the beginning of 1984, after her trade unionist husband, Fernando, was kidnapped and disappeared on his way home from work. Nineth started searching for him, placing adverts in papers and going to morgues. Along the way she met others who had lost relatives in the same way, and realised they could be more effective as a group. Initially they took a gentle approach: adverts, petitions and meetings with government representatives. A march of 1,000 people in October 1984 attracted international attention and led to the establishment of a Government Commission to investigate. But when this failed to produce any results, GAM began to use more assertive nonviolent tactics, such as the occupation of public buildings. By spring 1985, President Mejía Victores was accusing GAM of being 'directed by subversives' and Nineth and other leaders were getting death threats. Then in April 1985, two leading members of GAM, Hector Gomez and Rosario Godoy de Cuevas, were tortured and assassinated. Rosario's body was found with bite marks on her breasts and signs of rape. Alongside were her 21-year-old brother and her three-year-old son, whose fingernails had been pulled out. It was after this that Nineth turned to PBI to ask if they could escort her.

It feels like Guatemalans are challenging the same Cold War mindset human rights activists face in Eastern Europe, where groups like Solidarity are accused of being sponsored by the CIA. Meanwhile at home in Britain, CND and Greenham women are labelled KGB agents/communists. Basically everyone not on the 'government' side, is regarded as a subversive by the other. Our struggles for peace and human rights are connected, except the cost of opposition in Guatemala is far higher than at home and I do want to help.

But I had not expected to actually escort human rights activists myself. The US office of PBI had told me my Spanish was too poor.

'That won't matter at all, you just have to stay with them all the time, you don't need to talk', Alice tells me.

'What happens if some military guy or policeman starts yelling at us?'

DIARY: GUATEMALA CITY, 1985

'Your job is to let us know what is happening.'

The central idea is that if the GAM activists are accompanied by international volunteers, (a) it will deter kidnappers, or (b) in the worst case, when a kidnapping does occur, the international observer will find a call box and alert the PBI team. They will then make a fuss to officials, embassies and the press, in the hope of preventing the kidnapping from becoming murder. I am told I will accompany three women, Nineth, her friend Isabella, a sociology student whose husband was also kidnapped, and Genara, an Indigenous woman from Alta Verapaz whose brother, uncle and cousin have all been kidnapped. I am given a briefing sheet and a Peace Brigades ID card and told to attach a photo, then sent to sleep in a barrack-like bedroom with three mattresses on the floor and a curiously luxurious tiled bathroom.

The briefing sheet tells me to keep an eye out for white vans and military vehicles following us; to sit where I can see the doors in restaurants; to always walk on the roadside when accompanying someone along the pavement, so they cannot be snatched away by a passing vehicle; and to be aware that my 24-hour presence can be stressful for the activist, so to stay in the background. I'm not quite sure how I will manage those last two at the same time. The briefing does not deal with the one situation that really scares me. What if I am kidnapped myself? Images of the four American churchwomen killed, raped and murdered in 1980 in El Salvador flit through my mind. Then I think about what Nineth and her friends face every day and I am ashamed. I take the tiny Guatemalan worry doll that a friend gave me before I flew here. Apparently if you put it under your pillow at night, she does your worrying for you. I sleep.

'Kidnapping of escorts has never happened and is really unlikely,' Barbara, an English volunteer tells me at breakfast. 'But I will give you some Peace Brigades' cards to scatter on the ground in the unlikely event that it does.'

'Do you think anyone will pick them up?' I ask. 'Might they not be too scared?'

SORRY FOR THE INCONVENIENCE

Barbara does not answer. Genara just called to say she is fright-
ened to leave home because there are military folk hanging round
outside her door, and can Barbara come at once. She leaves without
giving me any cards. But it doesn't matter. When I arrive at Nineth's
to take over from Seb, a short-haired Canadian, he briefs me again.

'Jump up and down and shout "sequestration" if anyone is kid-
napped, make sure you have ten centavos for the phone, and throw
Peace Brigades' cards around.' He gives me his last remaining card.

Over the next two weeks I become completely absorbed in this new
form of nonviolent action, simply accompanying and supporting those
who are putting their lives on the line for the sake of justice for those
they love. I learn to say 'kidnapped' not 'disappeared', 'because these
people have not disappeared, someone knows where they are, there are
always witnesses', Alain, the priest who heads the Guatemala PBI
team, tells me.

Most of my time is spent with Genara, who lives with her two
small children in one room around a courtyard with a sink and loo
shared with seven other families. We take it in turn to wash from
buckets. On day shifts I accompany her and her children to the rows
of stalls selling traditional clothes at the airport, and other street
markets. Genara sells the silver, and beautiful hand-woven embroi-
dered materials made by her family in Tactic, to shop owners who
then make it into Western-style tops and dresses or handbags. On one
occasion I lose her among the market stalls, and she roars with laugh-
ter when she sees the panic on my face when I catch up with her. On
night shift she insists I have one of the three metal-frame beds in the
room, while one of the family takes a reed mat on the floor.

Everyone knows her and likes her. She was a health promoter in her
village and her brother was a primary school teacher. Her 48-year-old
uncle was also a health promoter and father of five; her cousin was a
cooperative worker and father of two. The only occasion I have seen
her cry is when she agreed to tell me her story. 'They hate us if we try
to advance ourselves in any way, politically, socially, economically. I

had to leave because they wanted to capture me as well, so I could not work in Tactic and the children were frightened.' She had not witnessed the kidnapping, but the soldiers had hit her mother and younger sister with their guns, so she knew who they were. 'Many in Tactic know but they don't say anything because they are afraid. The military killed 200 people and captured 100 in a little village nearby.' After she left, her family continued to search hospitals and morgues: 'The soldiers just laughed.' She heard of GAM through the newspaper adverts for the missing: 'I recognised in those adverts the same anguish that I felt.'

She told me she had no fear for herself 'because the pain I feel for my brother is all I can feel, there is no room for anything else':

> *I am angry, I respected the military before and thought they were doing a good job. My eyes have been opened. I had the illusion that I was in the process of self-development, that we were progressing to a better future, now I don't believe in that. I believe in GAM. The struggle is very important so that others don't suffer in the same way. It's important because the children have realised there are no human rights in Guatemala, that the military government cannot be trusted.*

In spite of the murders of their comrades she and all of GAM were completely committed to nonviolence:

> *We are insulted. The officials who rule cannot understand human suffering but that just makes us stronger. Our actions have made us like family, nonviolent but firm, strong and decided. No one has ever advocated violence; we don't want that as we would be compared with the military. There are stronger actions we can take but they will be pacifist.*

And then she thanks me: 'No one believes you influence us, but without the Peace Brigades it would be much more dangerous for us. We would continue our work but we would be more exposed, so thank you.'

I am completely humbled. Listening to her I realise why this sort of nonviolent action matters. Connecting, escorting, witnessing, advocating, is the least I can do to support her search. My risk is minimal and I can walk away. Genara cannot.

In my last week, I join the remaining PBI volunteers accompanying GAM on their first demonstration since they occupied the Cathedral. That occupation led to the deportation of PBI volunteers who were quite wrongly accused of 'influencing'. It had lasted a week and ended with the creation of a mediation team including the Archbishop, but Genara told me they achieved nothing. This time GAM held a simple vigil outside the National Palace. They gathered in the Plaza de la Constitución, and each person wore a straw hat with the name of the kidnapped loved one written upon it. Then they walked across to the National Palace holding up lists of the 800 names that Genara had spent the previous days writing out. They called out: 'Where are they now? Where are they now?' As previously, PBI took no part in the planning. Now we simply stand silent, keeping a good distance from the demonstrators, to witness. I watch Genara marking up posters, encouraging and helping others, and am once again impressed at her courage and humour. Then along with all the others she throws carnations at the door of the palace, shouting out: 'Before you change power, tell us where our loved ones are!'

8

EVERYTHING IS CONNECTED

'That is not what would happen today', Laura Gomariz told me
when I shared some of this story. We were sitting in the PBI
office in Guatemala City in October 2022. Laura is the training
officer, having first come to Guatemala as a volunteer in 2017.
These days volunteers must commit for a year, be fluent in
Spanish, and go through an intensive training and selection process prior to travelling to Guatemala. They also, as Laura
detailed, need to be able to handle living with other people in a
team house, look out for each other's wellbeing, and be able to
take and give feedback in a supportive manner. There are codes
of conduct to be signed, including on sexual harassment. There
is a friendliness and a professionalism that is very different from
my experience in 1985. Volunteers need previous experience in
some form of grassroots work, as well as the capacity to 'work
within the principles of non-partisanship, non-violence and non-
interference, and to work in a horizontal structure based on
consensus decision-making'.

The accompaniment is rarely for twenty-four hours and never
occurs when the individual or group is involved in illegal activity

or any form of civil disobedience. PBI wants to demonstrate its complete respect for the law in Guatemala: 'If it's a peaceful legal demonstration, no problem, but if it turns violent or there is any law-breaking, we leave.' For PBI, nonviolence in this context is not about direct action, it is about rejecting the use of violence as a means of resolving conflict, and 'promoting cooperation between groups that work in a democratic way and strive to find political solutions to conflicts through non-violent means'.[1]

PBI Guatemala is part of the international organisation that was founded in 1981, with the idea of sending international volunteers into areas of high tension in the hope that they might help to avert conflict. As it says on its international website, its aim is to 'provide protection, support and recognition to local human rights defenders who work in areas of repression and conflict and have requested our support'. Currently there are country projects in Colombia, Honduras, Indonesia, Kenya, Mexico and Costa Rica (for exiled Nicaraguan human rights defenders) as well as Guatemala. These field projects are supported by country groups in thirteen countries.

In fact, the Guatemala project closed in 1999. The Peace Accords of 1996 that brought an end to more than thirty years of internal armed conflict, generated hope of new democratic beginnings. But as the human rights situation deteriorated over the next few years, various organisations asked PBI to return, and the office was re-established in 2003. Today, as Silvia Weber, the communications officer, explained to me, a team of eight people focuses on accompaniment and advocacy in three main areas: those continuing the struggle begun by GAM against impunity and for transitional justice; those working against unjust land distribution and forced evictions; and the predominantly Indigenous organisations working to defend their territory and the environment from the depredations caused by extractivist projects by both national and international private companies.

Currently, for example, PBI have been accompanying activists resisting the expansion of large-scale sugar plantations on the Pacific coast; mining operations implemented by transnational corporations in the south and east; and the imposition of hydro-electric power (HEP) plants in the centre of the country.[2]

Asmamaw and I were doing voluntary work in Guatemala. We hoped we might use the opportunity to connect with some of these groups and learn from them. Every lecture I have given on the 'Climate and Ecological Crises' has emphasised that this is not just a climate emergency but a global environmental emergency, in which so many of the products we take for granted at home—minerals, bananas, sugar, to name just a few—are ripped out of countries where local communities have had no say as to how their land is used. As Greenpeace stated in a recent report, 'Black people, Indigenous Peoples and people of colour across the globe bear the brunt of an environmental emergency that, for the most part, they did not create. Yet their struggles have repeatedly been ignored by those in positions of power.' In challenging the systemic racism that exists in the Global North, including within environmental movements, they call for greater recognition and acknowledgement of the environmental work those groups are doing.[3] We weren't in a position to volunteer for PBI, but witnessing, learning and advocating for such groups appeared to us to be a good way of showing solidarity. Moreover, nonviolent climate activists from the Global North have much to learn from the Global South, both from their understanding of the issues and their strategies of nonviolent resistance.

The IPCC Sixth Assessment Report (AR6), published in April 2022, pointed out that 'Indigenous Peoples have been faced with adaptation challenges for centuries and have developed strategies for resilience in changing environments that can enrich and strengthen current and future adaptation efforts.' And on the 'International Day of the World's Indigenous Peoples',

SORRY FOR THE INCONVENIENCE

9 August 2022, the Secretariat of the United Nations Framework Convention on Climate Change (UNFCCC) stated:

> Through generations of close interactions with the environment, indigenous peoples safeguard an estimated 80% of the world's remaining biodiversity. Together, the global community has an opportunity to reorient the way it interacts with nature and build resilience for all through collaborating with and learning from indigenous peoples, the stewards of nature.[4]

But that collaboration does not appear to be happening in Guatemala, even though the three Indigenous groups, Maya, Xinka and Garifuna, make up 43.6 per cent of the population.[5] To understand why needs some understanding of Guatemala's past and present history. A good place to begin is the Plaza de la Constitución. Asmamaw and I walked there after our meeting with PBI. I wanted him to see the twelve pillars that hold up the wrought-iron fence in front of the Metropolitan Cathedral. Each one is carved with long lists of names, listed under the districts where they lived. These are memorials for at least some of those who were murdered or kidnapped during Guatemala's thirty-six years of internal armed conflict. Some 200,000 Guatemalans are thought to have been killed, and as many as 40,000 disappeared. Government forces carried out 93 per cent of civilian executions.[6] The majority of those killed and listed on these pillars were from the Indigenous population. Between half a million and 1.5 million are thought to have been internally displaced or forced out of the country.[7]

Guatemala is both heart-breaking and inspiring. Heart-breaking, because that war was just one episode in some 500 years of colonisation, expropriation, extraction and murder, beginning in 1524 when the Spanish arrived. Inspiring, because Guatemalans have repeatedly used many of the nonviolent strategies discussed in this book to confront, and at times throw out, brutal dictatorships. Hitler-admirer and fascist Jorge Ubico, who took power in 1931 with the support of the United States and the United Fruit Company, one of the main landowners in Guatemala, was forced out of office by a general strike, a month-long economic shut down in 1944.[8]

Nonviolence prevailed again in 2015, when an unprecedented leaderless coalition came together to demand the impeachment and resignation of both President Otto Perez Molina and vice president Roxana Baldetti, after it was revealed that they and other members of the government were involved in large-scale corruption. In weekly demonstrations, thousands of students, housewives, workers and business people gathered in Plaza de la Constitución. This culminated in a national strike in late August 2015, during which Indigenous and rural organisations set up road blocks across the country, while urban protests and marches took place in numerous cities. The president resigned, and both he and the previously arrested vice president were charged, although they were not finally convicted until December 2022, when both were sentenced to sixteen years in prison.[9]

The investigation and revelations came from Guatemala's Attorney General, Thelma Aldana, and CICIG, the International Commission against Impunity in Guatemala, an independent body established by treaty between the government and the UN in 2006 to help the Attorney General and state institutions confront corruption and organised crime.[10] CICIG was an effective and vital institution in the struggle to re-establish justice and the rule of law in Guatemala, according to Carlos Juárez, a lawyer who has worked for GAM since. We went to meet him in the GAM office, not far from the Cathedral. 'We got the opportunity to find justice for so many families in Guatemala', he told us. This included the trial and jailing of four people for forty years each, for the disappearance of Nineth's husband Fernando García,[11] and the trial and conviction of one of those involved in the killing of Genara's uncle.[12]

But the right-wing populist comedian Jimmy Morales, who was elected at the beginning of 2016 on the slogan of 'Neither corrupt nor a thief', shut down CICIG after it started to investigate his family finances, along with those of his party and the economic elite. He told the UN General Assembly in 2018 that CICIG was a 'threat to peace' and sowing 'judicial terror' in Guatemala. CICIG's mandate was not renewed and it closed in 2019.[13]

Today Guatemala is ruled by what both Silvia at PBI and Carlos Juárez at GAM identified as 'the Corrupt Pact', in which the three supposedly separate and independent arms of government—the judiciary, Congress and the president—now work together to attack and undermine organisations within civil society that work against corruption, for human rights and transitional justice, and to defend the land. Put simply, the checks and balances that sustain democracy have disappeared. According to Human Rights Watch, in June 2022, 'the Inter-American Commission on Human Rights included Guatemala, along with Venezuela, Cuba, and Nicaragua, in the chapter of its annual report reserved for states with grave, massive, or systematic violations of human rights and serious attacks to democratic institutions'.[14] Guatemala wanted to leave this Commission, as well as the Inter-American Court of Human Rights, Carlos told me.

The problem nonviolent activists face these days in Guatemala is no longer being kidnapped off the street by people in a white van—although eleven human rights defenders were killed in 2021.[15] The greater risk is 'criminalisation'. Activists who are seen to be involved in a specific cause, find they are charged post hoc for spurious offences supposedly committed years earlier.

Meanwhile, 'the prosecutors and judges who pursue human rights or corruption cases, face criminalisation, for that support'. Since 2018, at least twenty-two Guatemalan judges and anti-corruption prosecutors had gone into exile. Others were in jail without trial, on spurious charges.[16] The independent media was also under attack. José Rubén Zamora, the owner of *El Periódico*, known for investigating government corruption, was on trial for money laundering.[17]

'Why don't people demonstrate like they did in 2015?' I asked Carlos.

'Public demonstrations are impossible,' he replied. 'Nobody wants to do that; you are risking your freedom. They are going to see you, they won't arrest you for being on a march, but they are going to look for a way to create a crime just to put you in jail.'

GAM itself was being accused of collaborating with judges and prosecutors to get convictions and had been threatened with a number of charges. 'If they don't put us in jail as individuals, they are trying to close down the structures in which we work. The economic powers in Guatemala don't want organisations like GAM to keep working in human rights because they have so much responsibility for everything that happened in the past and for what is going on in the present.'

'Are you afraid?'

'Yes', Carlos replied, but he kept in mind what GAM Director, Mario Polanco, said to all of them: 'We worked in a really dangerous time where we could lose our lives. Right now, you can lose your freedom. I don't think it's an improvement but what matters is that we believe in what we are doing and we will keep trying.'

'Do you think this is the way change will be achieved? Through the law?' I asked.

'Of course. If you don't do anything, nothing is going to happen.'[18]

Back in our lodgings, I learnt from the BBC that at home in the UK, the new Public Order Bill had passed its third reading in the House of Commons, which would allow me to be tagged and criminalised before I had committed any crime. It seemed trivial compared to what Carlos, Bernardo Caal Xol and others faced here, but Guatemala shines a light on what happens when democracy is captured by a corrupt elite, indebted to oligarchic benefactors, that work to remove the checks and balances on power. What Guatemala was also teaching me was an even stronger version of what I learnt from Tim Crosland and Melinda Janki. Nonviolent direct action is not only about disruption and breaking the law to draw attention to an injustice. It is also about fighting to maintain and uphold the law and justice in the face of corruption and collusion. But maintaining and upholding this in Guatemala requires astonishing resilience and courage.

* * *

One example of such courage is Bernardo Caal Xol, a Maya Q'eqchi' teacher, trade union organiser and environmental defender from Alta Verapaz. In 2015 he discovered that hydroelectric plants were being constructed on the Cahabón river that would divert the water, cause deforestation and damage local communities. Further investigation showed that the companies responsible, OXEC S.A. and OXEC II S.A., had begun construction without consulting those affected, as required by both national and international law. Bernardo led the legal challenge against these companies. A year later, in February 2018, he was arrested for allegedly detaining and stealing from some OXEC workers in 2015. Laura from PBI accompanied him on the day he voluntarily presented himself in court. He was sentenced to seven years in jail, on the word of the OXEC witnesses. Amnesty International made him a prisoner of conscience. He was released in March 2022 after four years in jail, not because the Guatemalan state had acknowledged any injustice on their part, but for good behaviour. And while he continued to organise nonviolent resistance in Cahabón, he now faced other equally spurious charges.[19]

Silvia gave him our number and he called to say he could stop in Antigua on his way to Guatemala City, where he was meeting with lawyers to discuss the new charges. He was now accused of drawing a salary as a teacher when he was not teaching, even though the law allowed properly elected trade union organisers to do exactly that. Prior to meeting him I immersed myself in various interviews Bernardo had given over the years.

What impressed me in all of them was his stoicism and resilience in the face of injustice, in spite of a social media campaign of vilification and defamation. In one interview he explained the paradox of knowing he was innocent of the charges against him. On the one hand it was a comfort, on the other it made things worse, because he really had no idea how long the punishment could continue. If they could criminalise you once for something you had not done, they could repeat it at any time, so when would it end? And he explained the connection between their struggle to protect water and the climate crisis, pointing out that Guatemala was already affected by both devastating tropical storms that washed away people's homes, crops and belongings, and drought that could lead to fights over a drop of water. He argued for a water law that would protect water for everyone:

> Water is used twenty-four hours a day so that we have health and life, therefore it is a human right and should not be a commodity for a few who have monopolised the water. [...] Water is our life, and if we don't have water we will suffer or we will leave.[20]

Bernardo explained how he came to choose this work when we met. His parents were illiterate agricultural workers and his father wanted something different for his son. There was no school in the village so he lived with friends a ten-hour walk from home. After training as a teacher, he returned to his own village, where there was still no school: 'So I gathered the children and gave them classes for free. Everyone wanted to study.' There was also 'no nurse, no road, no drinking water, no light'. He gained his teaching licence in 1995 and started to mobilise in the surrounding communities to obtain these things: 'We succeeded, we got roads, education, schools and drinking water, from the government.'

In 2006, his colleagues elected him as leader of the teachers' union for all seventeen municipalities in Alta Verapaz, including Cobán. This meant giving up direct teaching in school and becoming a full-time organiser for the region. It was this that led to his accidental discovery of the hydroelectric projects. Driving his motorcycle down a road he had not been down for a long time, he noticed that familiar hills had disappeared completely and machinery and construction had appeared in the river. He asked local people what was going on but 'no one knew'. So he gathered representatives from the affected communities to form a commission and to get answers. The mayor and the municipality told them nothing; but then, in the Ministry of Energy and Mines they discovered the licences allowing companies to construct hydroelectric plants along the Cahabón river and its tributaries, affecting some 50 kilometres of river and more than 200 communities.[21]

Back in 1996, after the Peace Accords had been signed, the International Labour Organization (ILO) confirmed Guatemala's ratification of the ILO Indigenous and Tribal Peoples Convention 1989 (No. 169). As the ILO said in its press release at the time:

> The Convention obliges ratifying governments to respect the traditional values of tribal and indigenous peoples and to consult with them on decisions affecting their economic or social development. It also requires governments to respect the land rights of tribal and indigenous peoples.[22]

No one had been consulted regarding these HEP schemes, so Guatemala had broken the law. Bernardo filed injunctions, and the courts confirmed that the communities' rights had been violated. OXEC S.A. was temporarily ordered to suspend construction and to organise a proper community consultation. When it failed to do so, Bernardo organised one in August 2017, where 26,526 people voted against the projects and only 11 people supported them.[23] This result was ignored, and Bernardo was jailed the following year. As he explained to me:

> That's why they ordered my capture. While I was in prison, they took people from these eleven communities around OXEC, put them in a hotel, a government representative came, they simulated a consultation, signed it and took that to the court. It was totally illegal. They give 150,000 quetzales to each community every year, so these eleven communities love the project and defend it, but there are 195 communities surrounding this area and all of them are against the project.

At the time of writing, two projects had been completed with seven hydroelectric plants in operation. Two more were under construction. The Peaceful Resistance of Cahabón planned to take the struggle to the Inter-American Court.

Bernardo talked a bit about his experience in jail. He saw some difference between past and present: 'In the past you disappeared and they killed you; now they torture you with prison, and the impact on the family is severe.' His wife worked as a teacher to sustain the family and she and his two daughters campaigned with many others for his release, but he missed them when they were growing up. 'I am very indignant,' he told me, when I asked what kept him going:

> This keeps me fighting. My grandparents were slaves in those coffee fincas. They died deprived of their rights. My father died without knowing how to read or write, my mother never went to school and also cannot read or write. But now I have political knowledge. What is happening to us is planned, not accidental, and it is not from God, it is from the authorities. They keep us poor, and our families. This is the plan of just a few powerful people who control the state; they began in colonial times 500 years ago. If I feel afraid, that gives power to those people.

I told Bernardo that some people might ask why he was fighting 'green power' when he wanted electric light in their communities. He pulled out a wodge of pictures showing the devastation caused by the schemes: cleared deforested land, large concrete canalisations and completely dried-up streams. 'These companies are killing the river completely, killing the life there', he said, explaining how the canalisation into closed channels, along with the formation of artificial lakes and the placing of turbines, had devastated animal and plant life in the area. The Peaceful Resistance had also denounced the mass cutting-down of trees. The construction and security traffic had further damaged the surroundings, driving animals and birds away:

> The river has followed this course for 1,000 years. It had a natural course, this is dried up, and people who live by the old ways, what happens to them? When you go to the river to pick up a stone—how many animals do you find under that stone? And when it has dried? The shrimp, the fish, the snails, they are all dead. The theory is that after water has been used to hit turbines, it will be returned to the river. But what happens to the people who are affected by the dam? The people who wash their clothes, who use the water for drinking, who fish there? And what happens to the vegetation, the birds and the animals who come to feed there? They have no water.[24]

José Bo, Bernardo's colleague, made the same point: 'We used the water for fishing, medicinal plants, bathing. They diverted it into a tunnel and it disappeared.' He was quite clear the projects were neither green nor clean. He believed the diversions, dams and the building of access roads had increased the risk of floods and landslides. Moreover, plastic waste accumulated and caused blockages. Bernardo had arranged for his sister, Maria, to take us to one of the villages in the Cahabón catchment area, so we were now José's paying guests. It was early evening and we sat on the small terrace watching the light fade above wooded mountains. José's cultivated fields, growing corn, cardamon and cacao, stretched around us. He was proud that he grew enough to feed his extended family, including a delightful infant grandson and two young people he supported: 'In other countries, people buy their food in a shop, here we get it from nature: plant, grow, to the table.' Two young women had just taken corn to be ground at the communal mill.

José was on the elected executive committee of the Peaceful Resistance of Cahabón and had been involved since the companies had first arrived. He helped organise the community consultation. He systematically listed the reasons why they would not give up the struggle:

> First, because they are stealing. This land does not belong to OXEC, it belongs to the people, we have legal title, there are plans dating back 140 years and we want it back. When our grandfathers registered this land, they said it could be shared but not sold. So when the companies bought it, they broke the law. Second, we were not consulted. Third, they took our water.

The water was both ritually significant, used for bathing before any ceremonial event, and essential in practical terms. Now there was either too much or too little. It was piped and inconsistent and he had difficulties growing his crops: 'When it rains the pipes are blocked with soil and in the dry season there is none. Fourth, it has created divisions in the community. In this village, fifty-four of the seventy-eight families support the Resistance, the others support the company.' We had sensed this as we were walking down the road, when there had been a jokey but tense exchange with a neighbouring family.

José's wife switched on a small electric light, powered by their solar panel. 'Life without electricity is healthy', she told me. José and his oldest son Freddy didn't completely agree. Light had its uses for studying at night, among other things, but they liked being self-sufficient. According to Freddy, '75 per cent of the village use solar panels.' 'I wish I could say the same about my village', I told him.

Freddy, as well as being a young father, was studying social work at university. He had been involved in the Resistance since he was fifteen. A science teacher at school taught him about the importance of the environment: 'She taught us what was in our surroundings, what was consumed, what got destroyed, the need to protect forests.' He was enthusiastic about the various demonstrations they had organised outside OXEC, the Court of Justice and the prison when Bernardo was in jail. They were completely nonviolent: 'We tell police and those affected how long we will be there.' When they blocked the gates outside OXEC, for example, 'No one was attacked, they just suspended work.' He thought this form of pressure was effective. These manifestations, often attended by thousands, were what contributed to the original suspension ordered by the Court of Justice, as well as to Bernardo's release. He recognised the risk of criminalisation, but contrary to what Carlos had told us, it had not stopped him: 'They want to silence us but it hasn't worked. We have just become stronger.'[25]

What I am learning from the Resistance is that we have to think about what we mean by 'green energy', where that comes from and at what cost. For example, a recent global study of hydropower showed that while reservoirs produce less CO_2, they emit significant amounts of methane, a potent greenhouse gas contributing to global warming. As John Harrison, professor at Washington State University's School of the Environment in Vancouver, and a co-author of the study explains, 'on a per mass basis, methane has a much stronger impact on climate than carbon dioxide does, [so] the reservoirs' net climate impact is increasing'.[26]

Energy solutions that are not solutions are not the only problem. PBI, along with more than 200 other groups and organisations, sent a letter to COP27 in Egypt in 2022, in which they pointed out that

> The profit driven extractive model which has underpinned the global energy model has not provided the economic benefits or development promised to many countries, and has entrenched existing inequalities, including around access to and ownership of energy, and gender inequality. It must be transformed. [...] Disregarding the rights of local communities and indigenous populations in the race to a decarbonized economy by 2050, in particular those impacted by the boom in the extraction of the minerals needed for the transition, and by land-intensive renewable energy projects, is short-sighted. It will result in numerous human rights violations and a failure of the responsibility of governments to protect human rights as established by the United Nations Guiding Principles on Business and Human Rights.

The letter lists 'at least 369 attacks on human rights, labour and environmental defenders around the world since 2015, including 98 killings, related to renewable energy projects, and 148 attacks, among them 13 killings, related to transition minerals mining'.[27]

As Bernardo explained in an interview following his release from jail in March 2022, the Resistance is not about just one river, but is part of a wider movement across the country and beyond:

> Since we were children we have been taught to love nature and take care of it, it is part of our memory. That is why we will always raise our voices in the face of threats to natural resources, and in this particular case, in defense of the Cahabón River. We only give continuity to the struggle of the people that has been going on for hundreds of years in the face of the plundering by those who established the State of Guatemala 200 years ago. In 1821 a State was delineated that, although in theory involves us, in practice it has only dispossessed us of our lands, impoverishing the native peoples. In the communities' territories, the State is totally absent, as there are no public services for people to develop. The State only appears

when it arrives to support the companies and authorizes them to appropriate what we have taken such great care of. The struggle of indigenous peoples will continue until the day we manage to free ourselves from oppression.[28]

It is oppression in which we are complicit. Put simply, just switching to an electric car, and continuing our current models of industrial development and consumption with some kind of 'green hat' on in the Global North, is killing people here in Guatemala and beyond. There has to be another kind of transformation.

* * *

That night we slept in hammocks in the sleeping room and the next morning all of us caught the pick-up truck back into Cahabón to attend the fortnightly Resistance meeting. This was held in the courtyard of Don Mario, the movement's spiritual guide. When we arrived, twenty-five men and women from the surrounding communities were already sitting in a large circle around a small stone-encircled flower bed planted with candles and smoking incense. On one wall hung the Maya flag: four triangles of colour, symbolising both the cardinal points of the compass and different aspects of the Maya cosmos as Don Mario explained to me: red for the sun, black for darkness, white to the north for air 'from where the wind comes', yellow to the south for 'where the air arrives', and in the centre, green and blue symbolising earth and sky.[29] Bernardo stood in front of the flag, speaking passionately: 'We are not going to be quiet; we will continue our resistance.'

The meeting felt both democratic and familiar; it might have been a People's Assembly at home. Women were present in equal numbers and contributed as much as the men. The agenda was wide-ranging, beginning with the practicalities of a trip to some ancient Maya temple sites. 'We need to go to historical places to understand what we did in the past,' Maria told me. 'This is ours, our ancestors did this; many of us have had no opportunity to study. This is a chance to learn, share and practise rituals.' There were updates from other meetings, including a recent UN gathering in Geneva to discuss other mega projects, as well as Cahabón, which had rejected Guatemalan government statements about representative consultation and called for the consultations to be done again. A major part of the Resistance meeting was given over to discussing whether and how the movement should engage in formal politics. A young lawyer, Paolino Temo, was there from Movimiento Semilla, a relatively new political party which, as he explained to Asmamaw and me later, was neither right nor left, but stood for the five principles of democracy, plurality, equality, respect for nature and active participation. Thelma Aldana had stood as its presidential candidate before being forced to flee the country. As Paolino explained, 'If we do nothing, they will take more land and expand their influence; we want to help fight this illegal invasion. We want candidates from the base to help us challenge these destructive companies and what they are doing, in a democratic way.'[30]

The Resistance faced the same questions and challenges that nonviolent activists do in many countries. Is it worth putting energy into formal politics? And if so, how do you engage with a politically divided opposition that has not found a way to collaborate in order to gain political power? Maria summarised the situation for me. The Resistance realised in 2019 that they needed to obtain political power 'to govern ourselves and prevent those people massacring the environment'. The Resistance asked her to stand in local elections because they wanted their voice heard. Initially she was uncertain. Her mother's view was that 'being a politician is meaningless, your brother is in jail'. But Bernardo encouraged her, saying it was up to her, that she had the capacity, that it was really important and there were no other women doing it. She was elected, and was now a councillor in the municipal authority. Her pragmatic approach reminded me of Jan Power back in Penzance. It was hard work and there were costs. A fellow councillor told communities that if they supported the Resistance, they would get no funding for schools or other projects, because the Resistance opposed them all. Maria's family were threatened by neighbours, so she took it up in a council meeting:

'I told the councillor involved: "You are hurting me and my family, causing psychological trauma and if anything happens it will be your responsibility. I am clearly against the HEP, but I support other projects, so respect my ideas and my position as a woman representing my people."' The mayor ordered the councillor to apologise, which he did. The school projects continued: 'It's calm, but I can't say its peaceful.'

As to which party to join, all the existing oppositional parties were underfunded and weak compared to those in power. One, for example, was only interested in getting the Resistance to support their chosen candidate: 'They wanted us to follow them and not to empower us. Semilla said they would support the representatives chosen by the Resistance.' These divisions in the opposition depressed Maria; she wanted them to work together. This sounded very familiar. She laughed when I told her about my own attempts to get cross-party cooperation back home, and my expulsion from the Greens for recommending tactical voting. 'Also, it's all men, Semilla is still men, we are proposing women.' Like Freddy, she preferred Resistance work, in spite of the risks.

Maria Caal Xol is as warm, intelligent, feisty and courageous as her brother. Bernardo's early activism had some costs for his family and herself. She was eight years old when he was peacefully campaigning for a school and road in his village. One night, when he was out at a meeting, armed men came to the house and took her mother, her 12-year-old sister and herself hostage, saying they would kill Bernardo and all of them when he returned. Someone alerted Bernardo and he came back with 400 other villagers. The armed men escaped and were never seen again, but Maria told me that the need to be constantly on guard meant she had no chance to enjoy her childhood. Bernardo told her to study, as it would help to protect her. She studied psychology, trained as a teacher, as a translator of her own language, and now has three children. The Resistance in all its forms is her full-time work.[31]

Before we left, we went up the Cahabón river to the natural pools at Semuc Champey. No HEP diversions here: the Guatemalan government realises that the water falling through a series of shallow aquamarine lakes is too valuable a tourist destination to destroy. It was quiet, with only a couple of local families bathing. I floated in the blue water and then swam up a narrow channel towards the larger falls, water pouring over the pool edge on one side, a high wooded cliff on the other. Hummingbirds flitted through blossom above my head and I thought about what Don Mario had told us the previous day, when he had arranged a special cleansing ceremony and, while preparing pieces of copal incense to burn, had talked about water:

> When we are in a river we are immersed in life, the stream pours off mother earth, furnishing us with life along either bank, attracting life towards it: birds, flowers, insects. It floods outwards to give life to others. We can immerse ourselves in it, wash our bodies, our clothes. It carries away the dirt and it is always giving, it sings.[32]

Immersing myself in a singing river was not something I would do in any local river at home, nor on the harbour beach in our village, after a summer of raw sewage dumping by our privatised water companies.[33] Bernardo is right. If water is to give us 'health and life' and is 'a human right', it cannot be privatised, monopolised or commodified in any country.

Then we returned to Antigua. The series of earthquakes in the eighteenth century that drove the colonial government to build a new capital city some 50 kilometres away, means that what remains here is a beautiful town of cobbled streets, small coloured houses, plazas with trees and fountains and baroque churches, convents and monasteries. The headless saints that fill niches in crumbling arches, and the altars carved with intricate traceries of flowers and birds, only add to the beauty, and underline the paradox of Guatemala's tourist industry, which celebrates both the beauty of the 'Mundo Maya' and the colonial world that crushed it.

We went to visit the museum in the colonnaded Palacio de los Capitanes Generales from which much of Central America was governed by Spain, until the particularly devastating earthquakes of 1773. President Alejandro Giammattei had just opened an exhibition on '3,000 Years of Guatemalan Culture'.[34] The exhibit ranged from Olmec and Maya sculptures, through colonial art and twentieth-century modernism to the present day. An introductory board asked in Spanish, K'iche' and English: 'Where does Guatemalan art come from?' and then provided an answer: It was rooted in both 'Western tradition' transferred by the Spaniards through 'the institution of the Catholic church', and the 'Guatemalan territory [which] existed before the arrival of European influences, a great artistic tradition with examples of Maya architecture, sculpture and painting almost three thousand years old.'

'Another way to exploit us', Bernardo had commented when he caught sight of the advertisements during his visit. He is right. The explanatory boards neither acknowledged nor apologised for the cost of that transfer from Europe, both to the Indigenous population and the landscape they inhabited. So it is not surprising that the racism, exploitation and profit-driven extraction referred to in that letter to COP27 continues to this day. Environmental injustice is deeply connected to colonialism. If we fail to understand, acknowledge and correct past mistakes, we will simply repeat them.

* * *

EVERYTHING IS CONNECTED

A mother rolled her laughing 3-year-old son across the floor mats. Then the small boy got up, as his mother, in her beautiful hand-woven skirt and huipil, lay down, and he rolled her back in the other direction. Asmamaw and I were explaining how these simple games helped the physical development of the child, and built empathy and cooperation. Running workshops promoting the physical and mental health of mothers and infants for Q'eqchi' women in Alta Verapaz was the other reason we were in Guatemala. One message we wanted to communicate was that showing love, playing and communicating with your baby from birth, is good for both parent and baby. Encouraging mothers to share and demonstrate how they do this, and supporting them in addressing the stresses that get in the way, is a significant form of nonviolent direct action. Loving, responsive, empathic parenting creates loving, responsive, empathic children. We need them. Everything is connected.

We were in Alta Verapaz at the invitation of Majawil Q'ij (New Awakening), a not-for-profit Indigenous women's association that works to empower women and to challenge violence against them.[35] I first met their president, Maria Morales, in 1989, when I was carrying out doctoral research on nonviolent activists and she was a leading member of the 'National Coordination of Widows of Guatemala' (Conavigua), that worked to promote the human rights of Maya women who had lost partners in the armed conflict.

Maria is another of my heroes. She did not go to school, but taught herself to read and write in both K'iche' and Spanish by listening to the hymns sung in the local Catholic church and working out their spelling on the page. With the support of a local catechist, she became involved in community work, but at the height of the violence her father insisted they move to the coast for safety. It was there that she learnt that the catechist, along with three other community leaders, had been murdered.

Her feeling of responsibility as a survivor pushed her to go back to the western highlands to work with Conavigua, although she was single and not a widow herself: 'I said, he's dead, but I'm here and I've got to carry on the work, because we've been doing it. I had to give the people encouragement.' At that time the army was using forced militarisation to oppose the guerrillas, compelling rural Indigenous men to join 'civil patrols' in their communities, do unpaid heavy manual labour and report 'subversive activities'. Maria, along with others in Conavigua and other groups, risked her life to educate communities on their constitutional right not to join these patrols, and encouraged a boycott of military service. One of the things that struck me then was her commitment to nonviolence. In spite of the violence inflicted on her neighbours, she had no desire for revenge against the soldiers who had harmed them: 'They're also our people from the same pueblos. Instead of helping people they're just killing their own people, but it's not because they want to do it: they're just like us, but they've been treated like animals, they haven't been treated like people—so all they've got is a great anger.'[36]

Majawil Q'ij participated in the peace negotiations advocating for Guatemala to be recognised as a multicultural and multi-ethnic country, in which the identity and rights of all four ethnic groups—Maya, Xinka, Garifuna and Ladina—were respected. 'Racism has prevented this being a reality,' Maria told me when we met this time. 'They don't accept us as human beings.' When we arrived in early October 2022, she was challenging this in Guatemala City, where Majawil Q'ij had helped organise the Third International Summit of Indigenous Women of Abya Yala (for many Indigenous groups, Abya Yala is the preferred name for the American continent). More than 500 women had gathered from across Guatemala and abroad. Now we were having dinner in her home town of Chichicastenango, after completing the mother-and-baby workshops. It was the Day of the Dead, and she and her son, a student at San Carlos University, had been cleaning her father's grave.

Maria gave me a now familiar analysis of Guatemala's present problems, in which the lack of democracy was intimately tied up with the destruction of the natural environment: 'What happened before was military persecution, today it comes from the government.' The worst thing in Guatemala today was that 'wealth is concentrated in the hands of a few. Now they go into communities to establish mines, and gain more land for monoculture with palms, coffee and sugar cane. They are taking more and more.' She tied this to what she saw as a 'supermarket' model of food sales that depended on chemical fertilisers. She felt Indigenous women's work in protecting and respecting the land was unrecognised. Majawil Q'ij promoted a different approach:

> Women take care of mother nature. As a woman, when I am culti-
> vating, I don't use fertilisers. I nurture mother nature so she nurtures
> its children. To recover from destruction takes a long time because
> of what we have lost. Better to remove the idea of the market and
> empower women to cultivate, share and sell the excess.

Maria called this 'community economy'. Majawil Q'ij worked on this approach at the grassroots within communities, and developed a module so that it could be taught on the radio. They connected this to campaigning on the climate crisis and recently collaborated to organise a seminar on 'the promotion and dissemination of ancestral knowledge for harmonious coexistence and climate change mitigation'.

Their other main focus was to protect and accompany women victims of violence. Guatemala has the highest rate of femicide in Central America. In 2022, 624 cases were reported, part of the 2,168 recorded since 2008. Of these, 71 per cent remain unpunished.[37] Majawil Q'ij members accompanied women and helped them in their pursuit of justice and safety.[38]

But the violence continues. On 12 October 2022, the last day of the Abya Yala conference, some ex-military men armed with machetes and sticks broke into the closing ceremony and started attacking women. No one was seriously hurt, but as the organisers pointed out in their press release: 'The intimidation and violence tactics perpetuated during the armed conflict against Indigenous populations and women continue to be reproduced to silence our voices and demands in the face of an absent State that does not guarantee wellbeing for our lives.' The attack terrified those present, bringing back vivid memories for those who had lived through the armed conflict. They demanded that those responsible should be punished, safe spaces created, and that attitudes and acts of hatred, racism and machismo be sanctioned.[39]

Majawil Q'ij make a direct connection between violence against women, patriarchy and violence against nature. Maria's words reminded me of a speech I had heard back in October 2019 during our first action with Extinction Rebellion. Jenny Harper Gow, co-founder of Mothers Rise Up, was standing in Russell Square one evening pointing out that we had 'collectively wronged each other by putting the individual, power and wealth above kindness, nurture, protection and caring', and explaining, as I had tried to do forty years earlier on that 'Women for Life on Earth' march to Greenham Common, that these were not 'feminine' but 'human qualities. We are all capable of mothering this Earth.' She continued:

> Our future depends not only on how we treat each other but also on our relationship with the natural world. Just as the abuse of power over other humans will end so will our abusive relationship with nature. We will no longer use our dominance and superiority to fight against it. We will no longer disregard the delicate balance of nature which is essential for our own survival. We will no longer use the violence of capitalism to dominate Indigenous communities [...] We

will ask to learn from the Indigenous communities' vast knowledge of how to live in unity with nature.[40]

* * *

I wish Jenny and Mothers Rise Up could meet Majawil Q'ij and the Tz'unun Ya' Community Collective. While we were in Chichicastenango, Bernardo forwarded a tweet. It showed a short piece of film in which Maya women and men were hurling plastic rubbish over the metal fencing of a skyscraper office building. The text said simply: 'Residents of the lake basin leave the garbage they have collected in front of the Chamber of Industry.' Further investigation on social media brought up a short film in which Tz'utujil women and men from San Pedro la Laguna, on the shores of Lake Atitlán, gathered and bagged up quantities of plastic rubbish from the lake waters and shores, transported it by boat and truck to Guatemala City, and then ceremoniously paraded with baskets of rubbish on their heads to the Chamber of Industry. They carried a banner saying: 'In our language Tz'utujil, the word plastic does not exist.'[41]

'We want to give them back what is theirs, we want to leave death here, their death', a woman said on the film. In an accompanying lecture, another woman explained how the rains of the recent Tropical Storm Julia had increased the 'trash, mostly single-use plastics', in the lake. She saw the degradation caused as directly connected 'to the relations of domination and the forms of violence characteristic of the capitalist, patriarchal and macho model':

> In a not too distant future most of this material will disintegrate into dangerous particles that will mix and become invisible in the waters, others will sink in the depths of the lake, while the rest will follow its direction to the banks of the different villages to drown the lives of the species that depend on the sacred Lake Atitlán [...] This scene becomes the mirror that reflects the inconsistencies of consumerism

imposed on us by this capitalist system and which leads us to the destruction of the resources provided by Mother Nature.[42]

The speaker was Nancy González, coordinator of the Tz'unun Ya' Community Collective which brought together Maya women from thirteen villages in the municipality of San Pedro la Laguna. We wanted to meet her, so we spent our last days in Guatemala on the shores of Lake Atitlán, or, as it is called in Tz'utujil: Qatee' Ya', meaning 'our mother lake'.

It was a quiet and sunny day, and local children played on the basketball court next to the municipal offices where Nancy said we could meet. There were large numbers of posters pinned to the office walls. Alongside a banner calling you to 'Get your free vaccination', one advertised a course on defending human rights. Another advertised the Guatamalan Foundation of Forensic Anthropology, which has done tireless work over the years investigating, exhuming and identifying the remains of victims of the armed conflict. A third cartoon simply stated: 'Democracy is justice, liberty, union, respect, equity and equality'. On the opposite walls there were large-scale murals. One showed Indigenous women carrying water bottles on their heads in front of the lake. The bottles all said: 'Don't Privatise the Water'.

These depicted a struggle that had begun in 2014, Nancy explained, when she arrived. That was when a group of vacation home owners, predominantly businessmen, formed an NGO called 'Amigos del Lago'. They planned to 'clean' the lake by installing a Mega-Recolector (mega-collector), a sub-aquatic tube stretching around the entire lake, to collect the sewage water from the municipalities and sell it on to companies that could use the waste. It would cost millions and municipalities would be expected to contribute financially. The Tz'unun Ya' Community Collective was formed in 2017 'to unite and challenge this project'. For one thing, most human sewage was not directly discharged into the lake, but managed by individual households. 'We need individual biodegradation for sewage, in coordination with the municipality. This was about privatisation of the water.'[43] The Collective petitioned the Constitutional Court to halt the project because Amigos del Lago had not consulted local people, as obliged to do under ILO Convention 169. Indeed, the contempt some of the Amigos felt for the process of consultation was shared in an interview with Al Jazeera, in which one man said: 'This is not very politically correct what I am going to say but I feel we're at the point where it's like "shut up, we need to do this [...] we will explain to you [the Indigenous communities] what we are doing, but get out of the way, we are doing this"'.[44] At the time of writing, the Mega-Recolector project was on hold.

Nancy grew up in San Pedro, the child of two teachers. She left to study law 'because as an Indigenous person we face injustice, too much injustice', but suspended her studies before completing her thesis because she thought it was more important to engage locally in civil society. For the last four years she had worked as a full-time unpaid coordinator of the Collective, while also being the mother of three small children and the wife of a human rights activist.

We walked down the hill in search of coffee, and passed a mural of Oliverio Castañeda de León, a student leader and civil rights activist assassinated in 1978, age 23. I told Nancy how surprised I was at the overt expression of politics in the community. She laughed. 'Here the past is not forgotten; what's forgotten is repeated.'

Nancy explained that the most concerning form of waste was solid waste, particularly plastics. Research had shown that the lake was highly contaminated with microplastic particles, silently damaging both human and animal health, so they had taken multiple forms of action. Women from the thirteen communities had been collecting plastic from the lake shores on a monthly basis for the last nine years. But Nancy was clear. It was not individuals who were destroying the environment with irresponsible consumption habits; the problem lay with the industries producing the waste. In 2016 the Collective successfully persuaded the municipality to ban single-use plastics. The business community immediately fought back, going to the Constitutional Court to prove this was unconstitutional. But 'the court said that the municipality had the authority to decide'. The ban stayed in place, and Nancy hoped they would now ban all plastic bottles.

Then there was the direct action in the capital. The political and symbolic act of returning the rubbish was significant. There was a large dump in the mountains above San Pedro, and 'public money is used to maintain that place, rather than education'. The Collective wanted industry to stop solving its waste problem by sacrificing increasing amounts of land, and take care of its own products without dumping them.[45]

I gazed up at those mountains as we took the boat back across the lake. The combination of smooth-sided volcanoes and the wide expanse of water, endlessly changing in hue through the play of light and cloud, captures the heart. It has also tempted endless interference. The president and dictator Rafael Carrera planned to drain the lake in 1862. Pan Am helped the government introduce black bass in 1958 by dropping it into the lake from aeroplanes. The idea was to attract sport fishermen. Instead the black bass decimated the local population of birds, insects and aquatic animals.[46] There are birds there now. A great white heron stood in the reedbeds as we pulled into the dock of our small guest house. I could see quantities of multi-coloured plastic particles floating around the bird, along with bottles and cans. Lake Qatee' Ya' was a living being with rights that the law should protect, Nancy argued in her lecture. Right now she was choking.

I am in awe of the courage of all of those I met in Guatemala. Spending time and talking with the Resistance, Majawil Q'ij and

Nancy also helped me realise a central issue in our struggle. If we are serious about the climate and ecological emergencies engulfing us, we have to make fundamental changes. We have to challenge the concept of 'sustainability' itself and think about what we are actually trying to sustain. Is it the current economic system with its pursuit of endless 'growth', in which nature is a commodity that possibly deserves protection when it furthers that model, but may legitimately be transformed or destroyed if it does not? Or is it about sustaining the natural world in which we are embedded and on which our lives depend? If so, we have to alter our relationship with the planet itself, so that we are no longer central, but part of the whole.

DIARY

MAY 2023, CORNWALL

'You know that's fake news, don't you?' a man shouts at me from the audience.

I am standing in the Cornish Bank, a community arts space in Falmouth, beginning a talk about the connections between the climate crisis and the refugee crisis to some fifty Just Stop Oil activists and others. They also want to hear my thoughts on nonviolent action, past and present, before hearing Roger Hallam discuss slow walking. So while I had certainly expected to debate the value of disruptive tactics of all kinds, I had not expected to be disrupted myself.

The slide that has upset the man so much is my illustration of 'Differences in global temperature and average atmospheric carbon dioxide 1880–2021' using data I have taken from the US-based National Oceanic and Atmospheric Administration (NOAA) Climate. gov site. It graphically shows how average temperatures have increased in the last sixty years in parallel with a steady rise in atmospheric CO_2. We are now at 420 ppm (the safe level is considered to be 350).[47]

I have not even got onto discussing the impacts of the worst drought on record in East Africa, now entering its sixth year. These include 4.6 million acutely malnourished children; more than 1 million displaced in Somalia;[48] and Oxfam's estimate, in 2022, that the combined effects of the climate crisis and conflict in this region kills one person every forty-eight seconds.[49] Nor have I had the chance to repeat yet again the most important point, that responsibility for this suffering does not lie with the majority of those living in the Global South, but *with the richest 1 per cent of the world's population who emit more than twice as much CO_2 as the bottom 50 per cent of the world.[50] Put simply, one return flight from London to Rome emits more CO_2 than my sister-in-law, living in Ethiopia, emits in one year.[51]*

The man stands up and continues to shout about fake news. When confronted by the pub owner politely asking him to sit down quietly or leave, the man threatens to choke the owner if he comes closer. He then climbs onto the platform to join me, stands in front of the slide, and tells us he has a right to disrupt, because that Just Stop Oil activist disrupted decent working-class people trying to enjoy the televised snooker world championships when he jumped on a snooker table and covered it with orange powder paint.

'I agree with you, and I want to thank you for using this kind of nonviolent tactic to interrupt my talk, rather than coming in with a gun and shooting me.'

There is a brief moment of connection as the man looks at me in surprise, claps his hands and is quiet. But when I ask him if we can arrange an organised debate at another time, so that I can give my prepared talk on refugees this evening, he says no, and when I try to continue the talk, he pulls out the plug from the projector and mic.

I am sure there is a military manual somewhere which states that on some occasions the best way to advance is to retreat, so that is what we did, into the garden, where I completed my talk as other kind friends blocked the man from following and shepherded him outside.

So do I have your attention? I know as a writer I am more likely to have grabbed it describing this scene, than by giving you an account of the four days I spent the previous week with around 70,000 others, picketing government departments and surrounding parliament in what felt more like a town festival than a protest. This followed Extinction Rebellion's shift of tactics in the New Year to 'prioritise attendance over arrest and relationships over roadblocks'.[52] Some 200 other organisations were out there including Greenpeace, Friends of the Earth, trade unions and community groups. Twenty-six health groups and organisations created a Planetary Health Hub in Old Palace Yard.

Over those four days, health workers I had never seen before, the general public and other activists mingled around our tents, joining in activities and listening to talks. Asmamaw and I were on a panel to discuss Climate Justice in the Global South and Refugees. We went on Chris Newman's health and climate walking tour. This combined pointing out local places of interest—such as 55 Tufton Street, home to the Global Warming Policy Foundation, which regularly pours out climate denial—with short briefs on topics like air conditioners, which are excellent at keeping those who can afford them cool indoors, but pour warm air into the streets outside, heating the surroundings and everyone out there by as much as 1.5°C. Shutters do a much better job at keeping all of us cool.[53]

Children drew postcards with messages to send to their MPs, and on two wet, chilly mornings we picketed the Department of Health, asking for a public health campaign on the climate and ecological emergencies. One of my colleagues, retired GP Stuart Drysdale, met a senior civil servant who told him they were reading the leaflet inside and 'there's lots of us agree it's the right way to go'.

But praise from the Metropolitan Police who said the absence of any arrests over four days 'showed the benefit of effective discussion between police and protest organisers to mitigate serious disruption', was of no interest to the media.[54] Nor was Hugh Brasher, the director of the London Marathon (which had coincided with one day of the action), who described cooperation with Extinction Rebellion as 'an incredibly positive experience [...] what we need to bring people together to solve the problems of the world'.[55]

Meanwhile a young man on a snooker table gets global media coverage for a number of days. So I am not surprised that fifty young people have come to listen to Roger Hallam talk about joining slow marches. Speaking on Zoom from a hospital bed, where he is recovering from a broken leg, Roger repeats the well-documented prediction of hundreds of millions being forcibly displaced over the next decade because of weather chaos caused by digging up fossil fuels. So people have 'an enormous responsibility in the Western world to enter into resistance at this time because it's our governments that are creating this hell for millions of innocent people'.

What slow marching does, Roger argues, is offer a method of resistance that is 'not performative protest in the sense of we're just going to London with a banner' but, by slowly walking along a road, creates disruption while being technically legal. This presents the authorities with a dilemma: allow it to continue or pull people off the road? Either approach will generate publicity. 'Disruption of the public space, roads, cultural events and such like is undeniably the mechanism through which you engage a population in moral debate about where they're going [...] And of course, lots of people say well that alienates people, and I would say, of course, that alienating people is always part of the process of political change.'

While we are talking in Falmouth, the Public Order Act is signed into law. So 'locking-on', that is, attaching yourself to another person or object, is now a criminal offence, and the police have additional powers to stop and search anyone 'without suspicion' in specific places at specific times. Meanwhile, Serious Disruption Prevention Orders (SDPOs) mean that people who have committed two previous protest-related offences can be prevented from protesting again.[56] *Apparently the police need these proactive powers to stop protests before they start. The message is clear: 'Don't even think about it.'*

The Home Office tweets that the new Act 'protects the fundamental right to protest, brings new penalties for disruptive and dangerous guerrilla tactics and reduces delays to the law-abiding majority'.[57]

The UN High Commissioner for Human Rights, Volker Türk doesn't agree. He urges the UK to reverse the legislation because it is incompatible with their international human rights obligations. He is particularly concerned that 'the law expands the powers of the police to stop and search individuals, including without suspicion;

defines some of the new criminal offences in a vague and overly broad manner; and imposes unnecessary and disproportionate criminal sanctions on people organizing or taking part in peaceful protests'. Moreover, it appears to specifically target nonviolent environmental and human rights activists: 'As the world faces the triple planetary crises of climate change, loss of biodiversity and pollution, governments should be protecting and facilitating peaceful protests on such existential topics, not hindering and blocking them.[58]

It is actually difficult to understand why the extra powers are necessary given the punitive ones already available under existing legislation. Roger tells us he has just spent four months in prison for conspiracy to cause a public nuisance, having been recorded in a previous talk calling on people to block the M25 and climb gantries. Morgan Trowland and Marcus Decker were recently sentenced to an unprecedented three years and two years and seven months respectively, for suspending themselves over the Queen Elizabeth Bridge at the Dartford Crossing last year. The judge told them it was necessary to deter copycat actions.[59]

We get the opportunity to see how the new rules apply on Coronation Day on 6 May. Instinctively republican, I am still curious about the ceremony, so Asmamaw and I catch up on highlights on the BBC on Saturday evening. There is something bizarre about watching the Archbishop hand Charles the ceremonial sword, telling him that it is a 'symbol not of judgment, but of justice; not of might, but of mercy' and that with it he should 'do justice [and] stop the growth of iniquity',[60] *while outside the police have arrested fifty-two people using their new powers.*

Those proactive arrests, made before anyone had committed a crime, include three volunteers who work for Westminster City Council and hand out personal alarms to vulnerable women at night; an architect and royal fan who simply wanted to watch the Coronation; nineteen JSO activists who hoped to reveal their T-shirts and hold up a banner, and were not carrying equipment of any kind;

Animal Rising activists holding a nonviolence training session miles from the Coronation area; and six members of the anti-monarchy group Republic, who had discussed their plan to have a rally and hold up banners and placards with the police for months in advance, but who had unwisely secured their banners with straps in a van.

When one of the JSO activists challenged a police officer about being arrested for wearing a T-shirt they were told that 'Public nuisance is anything the police say it is.'

The arrests are headline news over the following days with criticism from human rights organisations and questions from the Mayor of London, followed by a parliamentary select committee hearing. However, although this committee is taking evidence from Republic and other affected individuals, no one from Just Stop Oil is invited to be present, and when they attempt to read out a statement they are evicted from the chamber.[61]

Meanwhile, the evidence of looming and actual catastrophes piles up around us. I think about our political representatives sitting in their offices in Westminster, and wonder how so many of them manage to avoid seeing what is going on. They only have to look across the river to St Thomas' Hospital. It was one of two hospitals in London whose IT systems had a major meltdown in last summer's heatwave, so that operations and appointments had to be cancelled and patients diverted because there was no access to their notes.[62] This was the same heatwave that burnt down homes, buckled railway lines, melted runways and caused an additional 3,000 deaths in the UK.[63]

Or they can cast their minds a little further afield and recall how that same heatwave triggered raging wildfires, melted glaciers and caused 61,000 excess deaths across Western Europe.[64] Meanwhile, the worst drought in 500 years disrupted farming and food supplies, made it possible to walk across the Loire, and brought shipping to a halt on other major rivers. Further heatwaves were predicted in Europe for summer 2023.[65] People in South and South East Asia are already suffering a potentially lethal combination of record-breaking heat and air pollution,[66] while in Canada wildfires displaced at least 13,000 in spring 2023. As I write, more are predicted.[67]

And if they don't want to think about the weather, they could consider what's happening to wildlife. Apparently, we are doing a very good job at killing it off. Dr Yuangeng Huang, at the China University of Geosciences in Wuhan, who led a recent study on the subject, told The Guardian: 'We are currently losing species at a faster rate than in any of Earth's past extinction events. It is probable that we are in the first phase of another, more severe mass extinction. We cannot predict the tipping point that will send ecosystems into total collapse, but it is an inevitable outcome if we do not reverse biodiversity loss.'[68]

Or they could simply look down at the Thames, that tidal heart of London, and contemplate what is happening to our water. Our oceans, covering 71 per cent of the planet and essential to producing the oxygen we breathe, are acidifying, and the currents are shifting in ways that will profoundly destabilise both ecosystems and the climate.[69] By 2030, we will have 40 per cent less fresh water than we need. Why? Because, as the Global Commission on the Economics of Water makes clear, our global mismanagement of water 'has pushed the global water cycle out of balance for the first time in human history. We have breached the planetary boundaries for water that keep the Earth's system safe for humanity and all life. Dangerously, the water crisis is increasingly intertwined with global warming and the loss of biodiversity, with each reinforcing the other.'[70]

Given all of these problems and the predicted millions who will be on the move trying to escape them, perhaps they could also ask themselves why this country, along with the six other big greenhouse gas emitters, spends twice as much on policing its borders to keep people out, as it does on climate finance to help poor countries address the problems that force people to move in the first place.[71]

Do our representatives honestly think that trying to silence those shouting about all of this and demanding action, will make these problems disappear?

Perhaps Artificial Intelligence has the answers. Recently a team of research scientists decided to enhance ChatGPT by 'integrating the information from the Sixth Assessment Report of the Intergovernmental Panel on Climate Change (IPCC AR6), the most comprehensive, up-to-date, and reliable source in this domain', and they created ChatClimate.[72] *I asked it: 'What three things should activists do to halt climate change?' It answered in seconds:*

Based on the information provided, here are three things activists can do to halt climate change:

1. Advocate for radical climate action beyond emissions reductions through collective disruptive action and nonviolent resistance to promote awareness, a regenerative culture, and ethics of care.

2. Advocate for new narratives or framings for climate mitigation, criticize positive meanings associated with high emission technologies or practices, show disapproval for high-emission behaviors, model behavior change, demonstrate against extraction and use of fossil fuels, and aim to increase a sense of agency amongst certain social groups that structural change is possible.

3. Prioritize car-free mobility by walking and cycling and adoption of electric mobility, reduce air travel, heating and cooling set-point adjustments, reduce appliance use, shift to public transit, and shift consumption towards plant-based diets.[73]

9

LIFEBOATS

I live in Mousehole, a small seaside village at the far end of West Cornwall, where on 19 December 1981, eight crew members of the *Solomon Browne* lifeboat gave up their own lives to try and save those of eight complete strangers. I go past the old lifeboat station from which they launched every time I go to town.[1] It is now a listed building.

For more than 200 years, groups of people in coastal communities have gathered together without pay to risk their lives to save complete strangers from drowning at sea. Given a common problem, people will work out how to address it, through mutual cooperation, by trial and error, without the need for externally imposed authority. People can develop complex organisational structures that save lives, from the bottom up. I think a lot about lifeboats. They give me hope.

The empathy and humanity of those involved and the support that the Royal National Lifeboat Institution (RNLI) has from the public at large, was powerfully demonstrated in July 2021, when Nigel Farage publicly attacked it on GB News and Twitter, for acting as a 'taxi service for illegal trafficking gangs'. Donations to the RNLI went up by 2,000 per cent in twenty-four hours, followed by their best fundraising year since its foundation in 1824, and a doubling of their supporter base.[2]

RNLI chief executive Mark Dowie put out a statement in July 2021, making clear that international maritime law permitted and obligated them to 'enter all waters regardless of territories for search and rescue purposes'. Moreover,

> when it comes to rescuing those people attempting to cross the Channel, we do not question why they got into trouble, who they are or where they come from. All we need to know is that they need our help. Our crews do what they do because they believe that anyone can drown, but no one should.

As one crew member explained:

> My motivation is to stop anyone drowning and washing up on the beaches. I don't care what time of day or night it is, a life is a life, and I will continue to give my best to the RNLI to protect as many as we can. [...] I am grateful that the RNLI support us and that we don't discriminate against anyone. I am proud of the work that we do and the lives that we have saved. I want us to shout about what we do and the care and empathy that we show. This country is having a crisis of empathy and I love that the RNLI are standing up for our morals and showing what I truly believe is the Britain we should all be proud of.[3]

Anyone can drown, but no one should. This is not about globalism or nativism but humanity. What lifeboat volunteers demonstrate is our ability to empathise and care for others, whoever they are and wherever they are from, and that in the face of impending or actual disaster, we will reach out to save each other. It is the same empathic care for strangers that I have seen in every crisis in which I have worked, not only from volunteers and relief workers, but most notably from the victims themselves. During the Covid pandemic I heard of refugee women on

Lesvos in Greece making masks for local nurses, and refugees in Athens delivering food parcels.[4] I saw it in the networks of mutual support that developed in my own village and across the country at that time. Anyone can drown but no one should.

Shortly after the RNLI put out these compassionate statements in the summer of 2021, I went back to Greenham Common for the first time in almost thirty years. Theatre director Rebecca Mordan had organised a march from Cardiff to mark the fortieth anniversary of the original march, and I went to meet their arrival on Greenham Common, because that is what the site of the airbase had become. In one of those small ironies of history, when the MOD introduced new by-laws back in 1985, restricting movement in and around the base in order to curb protests and criminalise the women, two Greenham women took them to court. It was this five-year legal struggle, ending with the House of Lords declaring those by-laws invalid in 1990, which helped establish that the MOD structures as a whole were illegal because they contravened existing commoners' rights. Following the removal of the last missiles in 1991, the MOD attempted to extinguish those commoners' rights, but further legal challenges from both the Camp and existing commoners resulted in the MOD selling the base to the Greenham Common Trust in 1997 for £7 million. The buildings were turned into a small business park, while the open land of the airfield was sold to Newbury District Council for £1. By 2001, public access and commoners' rights had been re-established, and the last women had left. Today Greenham Common is a site of special scientific interest; ponies and cattle graze where there were once runways; and a commemorative peace garden has been created at Yellow gate where I once lived.[5]

The morning after meeting the women, I went for a walk across the common. Swallows flew in the air. A wren startled as I picked ripe blackberries. I passed thickets of young oak trees, walked through long grass past gorse bushes and met ponies grazing. The perimeter fence was long gone and I could not orient myself or recognise anything, until I reached a deep, reedy pool in front of grass-covered silos. 'There are bats roosting inside', a bird watcher sitting by the pool told me. The silos were still standing because Historic England had made them a Scheduled Monument, describing them as:

> internationally important as one of the key emblematic monuments of the Second Cold War, signifying an escalation of the nuclear arms race by the introduction of GLCM [Ground Launched Cruise Missiles]. Subsequent to the Intermediate-Range Nuclear Forces (INF) Treaty of 1987 most of the missiles and launchers were destroyed. [...] It remains a potent symbol of the positive power of arms control treaties to render advanced military technology obsolete.[6]

I agree. For me, the Common and those empty silos symbolise the possibility of transformation for the better. I walked back to join the crowds of Greenham women, guests and local people holding a small festival inside a large open marquee near the old Air Traffic Control Tower, now a heritage centre. Billy Drummond, Mayor of Newbury, was giving a welcoming speech to the marchers, telling them that he 'personally believed the camp contributed to the end of the Cold War'. Jeff Brooks, the deputy mayor of neighbouring Thatcham agreed, asking the assembled crowd if they remembered being asked to shelter under a table in the early 1980s:

> Let's be clear, the Western world has a lot of things to thank you for [...] The enduring protests, washing at standpipes, shunned by shopkeepers, targeted by police; not [for] two weeks in Trafalgar Square but nineteen years. The hardship you endured was immense. [...] You made it happen, before the Berlin Wall came down, and we can only thank you for that, all these years later.[7]

I took a bus tour to nearby Aldermaston where women have established a monthly peace camp. Rebecca Johnson, one of my contemporaries at Greenham, who co-founded the Aldermaston

Women's Camp in 1985, was acting as our tour guide. She told us about meeting Mikhail Gorbachev in 2004 at a Pugwash meeting in London, where he was the main speaker.[8] She recalled him being asked what made him trust Ronald Reagan and go to the Reykjavik Summit in 1986, which led to the INF Treaty being negotiated and signed in 1987. Gorbachev replied that he knew that he had to take the risk of going to Reykjavic, because it was the only way. It wasn't Reagan he felt he could trust but 'the Greenham Common women and the peace movements of Europe [...] they would not let America take advantage if we took this step forward'. Rebecca stood up, introduced herself as a Greenham woman and thanked him for allowing her to 'stop living in the mud outside the US nuclear base and to get on with the rest of my life'. After the talk she went up to say hello: 'He just held my hand very, very warmly, for several seconds, saying something like, I am very happy to meet a Greenham woman.'[9] He had made a similar point in a speech to the UN, saying that it was Greenham women who had prompted him to question the rationality of the arms race.[10]

'Are we safer now?' I asked Rebecca when we met up again on Zoom, shortly before the Extinction Rebellion actions in London in April 2023, where Greenham women, along with numerous other peace groups, were among the 200 groups surrounding parliament. 'If you think in terms of the numbers of nuclear weapons, we are far better than we were in the mid-1980s,' she replied, 'but that's still 13,000 too many for safety.' Many are more destructive than in the 1980s arsenal; over 92 per cent are divided between the US and Russia, with the rest held by seven other nuclear-armed states. Rebecca went on to remind me of the facts that framed my life in the 1980s and that remain unchanged. It would only take a small regional exchange using less than 1 per cent of the existing nuclear arsenal to cause a global nuclear winter, due to the dust clouds lifted high into the atmosphere which would block out the sun's rays, causing rapid freezing. Two billion could die in the famine that would follow.[11]

Meanwhile, the war in Ukraine brings home their uselessness and danger. An aggressive, nuclear-armed state has been unable to coerce a non-nuclear state into submission, and the risk of actual use in the face of defeat grows by the day, as does the ever-present risk of detonation by accident, as weapons remain on high alert.

Rebecca recognises how US-rooted 'military-industrial-bureaucratic-academic establishments' (MIBA) fuelled nuclear proliferation from the 1950s by branding nuclear weapons as 'deterrents' and dismissing disarmament as just 'an ideal to strive for but never achieve'.[12] She points to the parallels with the threat posed by the fossil fuel industry:

> There are two primary routes to total extinction of the human race, nuclear war and climate destruction, and their root causes are the same. Whether through fossil-fuelled greed, planetary overheating and the belief that we can control nature for human advancement— or through nuclear war, mass murder, radioactivity and the planetary freezing that leads to nuclear winter, we are staring extinction in the face unless we quickly change course.[13]

Yet lifeboats are out there, even if not everyone wants to climb on board yet. This is thanks in part to years of activism by people like Rebecca who, since leaving Greenham after going to prison a dozen times or more, has dedicated herself to research, strategising and global campaigning to abolish nuclear weapons. In 1995 she founded the Acronym Institute for Disarmament Diplomacy (AIDD); she has been vice chair on the board of the Bulletin of the Atomic Scientists, and in 2010 set up the Australian-founded International Campaign to Abolish Nuclear Weapons (ICAN) in Geneva, becoming its first president. In July 2017, the UN Treaty on the Prohibition of Nuclear Weapons (TPNW) was adopted by the majority of the world's nations. In the same year, ICAN was awarded the Nobel Peace Prize for 'its work to draw attention to the catastrophic humanitarian consequences of any use of nuclear weapons and for its groundbreaking efforts to achieve a treaty-based prohibition of such weapons'.[14]

The TPNW came into force in January 2021. What makes it different from other arms control agreements which seek to limit the numbers or types of nuclear weapons, is that the UN nuclear prohibition treaty's focus is on preventing the humanitarian catastrophe that would be caused by any use at all. In this way it resembles treaties prohibiting the use of chemical and biological weapons.[15]

Ninety-two countries have signed to date, including sixty-eight that have ratified, that is 48 per cent of all countries.[16] The fact that the nuclear states and their allies have not yet signed does not make it irrelevant, because, as Rebecca explained, as increasing numbers of countries join the treaty it 'embeds the norms', with prohibitions that are 'very clear and itemised', including 'all the activities, operational and physical, material, or financially funded activities that could enable anyone to acquire, use, or threaten to use, or deploy nuclear weapons'. 'Anyone' means the treaty prohibitions apply to states and to non-state actors, including companies.

This opens the door to creative transnational activism in supporting those goals. For example, ICAN has a project called 'Don't bank on the bomb' which has been very successful 'because banks are risk averse'. The campaign engages with banks and investors that have funding in both nuclear weapons development and in countries that are signatories to the TPNW, where they are subject to the law of that country. 'So they either have to pull their activities out of that country, or pull their activities out of nuclear weapons,' Rebecca explained: 'Guess which they're choosing to do?'[17] To date, 101 banks and financial institutions, including Deutsche Bank in Germany and the Co-operative Bank in the UK, have divested from companies involved in nuclear weapons, many citing the TPNW.[18]

Rebecca is not asking climate activists to refocus on nuclear weapons: 'I want them to carry on with exactly what they're doing. [...] Climate justice and nuclear abolition go hand in hand. It is only by campaigning for both of those things that we can build the structures of peace that will undermine the structures of militarism and industrial destructivism.'

That was why she was also working on building support for a Fossil Fuel Treaty. The treaty aims to put in place legal obligations to phase out fossil fuels 'rapidly enough to avoid catastrophic climate change, while ensuring equity and a global just transition'.[19] Rebecca argues that treaties aren't the only answer, but believes that while 'it's vitally important to raise public awareness with activism, [...] we also need to push achievable, transformative, political and legal demands onto the agenda of decision-makers in the UN and governments.'[20]

We need that global just transition. The growing awareness that environmental issues and demands for equity and global justice are deeply connected is another source of hope. We cannot carry on with business as usual. We know that it would take five planets for everyone to have the lifestyle of the average American consumer.[21] We know that we are already overshooting planetary boundaries, graphically illustrated by the fact that we now have more plastic on the planet than the mass of all land animals and marine creatures combined.[22]

Recycling plastics won't solve this. We now know all those plastic bottles and cartons we throw in the recycling bin are being turned into microplastic particles that are contaminating water and life across the planet.[23] Nor will switching over to non-fossil fuels and carrying on over-consuming as usual. The International Energy Agency (IEA) points out that:

> An energy system powered by clean energy technologies requires more materials than fossil fuel-based counterparts for construction. Minerals are a case in point. A typical electric car requires six times

the mineral inputs of a conventional car, and an onshore wind plant requires nine times more mineral resources than a gas-fired plant of the same capacity.[24]

But human beings are not insatiable consumers. A lifetime of work as a psychiatrist in multiple countries has made it clear to me that once basic needs are met and one has security, happiness does not come from the accumulation of material goods, bigger houses, more cars, clothes, shoes or ornaments. Rather it comes from social connections, friendships, opportunities to be kind, to be creative, to play, engage in sport, get close to nature, and from the feeling that one has some say and some control over what happens in one's own life and the lives of those one loves. None of these things depend on how many material goods an economy produces, or can be measured in terms of Gross Domestic Product (GDP). Yet our governments still appear to be stuck on a linear track, pushing the endless extraction, production and consumption of stuff to feed that metric, while telling us that the way to address the climate crisis is a simple switch to 'green growth'.

In my Ethiopian family home I witness the costs. The local systems of burying waste in pits have been overwhelmed. The streets are littered with non-compostable garbage. Spoonbills step between discarded plastic water bottles around the ever shrinking lake. Guatemalan friends gave me a glimpse of what continuing extractivism might cost, when global commons such as water are turned into commodities that can be sold. Back here in Britain we continue to learn what privatised water means: the degradation of essential infrastructure, toxic sewage on beaches and in rivers, all in the pursuit of profits for shareholders.

Yet even while most of our political representatives continue to trumpet the Goddess of Economic Growth, another language is emerging that challenges this. It is not a new one. Robert F. Kennedy, when running for president of the United States in

1968, made a speech to the University of Kansas that seems just as relevant to present times:

> Our Gross National Product, now, is over $800 billion dollars a year, but that Gross National Product [...] counts air pollution and cigarette advertising, and ambulances to clear our highways of carnage. It counts special locks for our doors and the jails for the people who break them. It counts the destruction of the redwood and the loss of our natural wonder in chaotic sprawl. It counts napalm and counts nuclear warheads and armored cars for the police to fight the riots in our cities. It counts Whitman's rifle and Speck's knife, and the television programs which glorify violence in order to sell toys to our children. Yet the gross national product does not allow for the health of our children, the quality of their education or the joy of their play. It does not include the beauty of our poetry or the strength of our marriages, the intelligence of our public debate or the integrity of our public officials. It measures neither our wit nor our courage, neither our wisdom nor our learning, neither our compassion nor our devotion to our country, it measures everything in short, except that which makes life worthwhile.[25]

Today, Guyana lawyer Melinda Janki is similarly frustrated: 'GDP growth is destruction. It's a Global Destructive Project. [...] When you say "I've made a million dollars out of forestry" what you've done is destroyed it, and you're not recording that destruction anywhere.' This was one reason she had introduced the concept of 'Natural Capital' into Guyanese law. This requires the Guyanese Environmental Protection Agency to give value to the nation's ecosystem, including air, water, wildlife and plants. Thus the profits made from clearcutting trees have to be weighed against the losses to the climate, species and people that benefit from their presence.[26] It is the same with fossil fuels. Melinda argues that if the deaths caused by air pollution and global warming are accounted for, then 'no fossil fuel company has ever made a profit because if you took into account the damage they've done to the Earth, they owe us. Big-time!'[27]

Jason Hickel, an economic anthropologist at the Autonomous University of Barcelona, advocates the idea of 'degrowth': 'This is not about depriving anyone of the things needed for a decent life. Our planet provides more than enough for all of us; the problem is that its resources are not equally distributed. We can improve people's lives right now simply by sharing what we already have more fairly, rather than plundering the Earth for more.'[28] What that requires is 'scaling down destructive sectors such as fossil fuels, mass-produced meat and dairy, fast fashion, advertising, cars and aviation, including private jets. At the same time, there is a need to end the planned obsolescence of products, lengthen their lifespans and reduce the purchasing power of the rich.'[29]

These ideas are creeping into the mainstream. Both the 2022 reports from the IPCC and the Intergovernmental Science-Policy Platform on Biodiversity and Ecosystem Services (IPBES) acknowledged the need to look at alternative models of development. In May 2023, the European Parliament hosted its second 'Beyond Growth' conference.[30]

Prior to this, some 400 civil society organisations and many of the economists attending the meeting published an open letter calling for a 'post growth Europe'. This would require 'a democratically planned and equitable downscaling of production and consumption, sometimes referred to as "degrowth", in those countries that overshoot their ecological resources. This is Europe's global peace project, because its current economic growth is causing conflicts both in and beyond Europe.' This could not be left to the markets: 'For the economy to serve the people, rather than the other way around, people must be given back control over the economy.' The letter ended with a call for 'beyond growth policies' based on the four principles of:

- *Biocapacity*: fossil fuel phase-outs, limits to raw material extraction and nature protection and restoration measures for healthy and resilient soils, forests, marine and other ecosystems [...]

- *Fairness*: fiscal instruments to foster a more equal society by eradicating income and wealth extremes, as well as super-profits [...]
- *Wellbeing for all*: secured access to essential infrastructures via an improved, ecologically-sensitive welfare state [...]
- *Active democracy*: citizen assemblies with mandates to formulate socially acceptable sufficiency strategies and strengthen policies based on ecological limits, fairness and wellbeing for all and a stronger role for trade unions [...].[31]

Unsurprisingly, neither the letter nor the conference received much coverage in the mainstream media. But the good news is that many of the changes which the letter called for in order to put those principles into practice, are already happening in different places around the world.

Take housing: in Vienna, a city of 1.8 million people, city-owned flats or cooperative apartments constitute half of the housing. More than 60 per cent of the city's inhabitants live in subsidised housing. According to Deputy Mayor Kathrin Gaal, 'Social housing policies in Vienna have been shaped by the political commitment that housing is a basic right.'[32]

It is also a basic right to benefit from the work you do, rather than be exploited to increase shareholder dividends. Cooperatives such as the Mondragon Corporation, founded by a Catholic priest in the Basque region of Spain in 1956, exemplify another approach, in which those who do the work make the decisions and share the profits. Mondragon now has more than 80,000 employees working in 100 different sub-cooperatives, involved in electronics, mechanical engineering, finance, retail and construction on five continents.[33] Moreover, it has its own university, where one research group is working to facilitate the transition to sustainable energy and the development of circular economies in businesses, industry and public institutions across Europe.[34]

As I write, the French government implements a ban on any short-haul flight in the country that can be made in less than two and a half hours on its publicly owned railways, as well as limiting the use of private jets for such journeys. The ban was first proposed by a Citizens' Assembly and integrated into France's 2021 Climate Law,[35] demonstrating once again that a combination of public ownership and citizen engagement in democratic governance produces good results.

The economists' letter recommends 'universal basic services (including the human rights to health, transport, care, housing, education and social protection etc)', and the evidence is that publicly owned services do make people happier. A cross-national comparison of industrial democracies in 2018, found that public employees are happier and exhibit greater life satisfaction than otherwise similar others, while subjective wellbeing varies positively with the size of the public sector.[36]

But how to pay for good public services? The answer is the old straightforward one: tax the rich. It's been done before with good effect. Republican President Dwight Eisenhower in the United States made those earning more than $200,000 ($2 million by today's standards) pay up to 91 per cent tax during his presidency in the 1950s. I grew up in a country where higher taxation was used to grow public services, before our own Mrs Thatcher told us 'there is no such thing as society' and 'people must look after themselves first'.[37]

Oxfam has called for immediate windfall taxes to address crisis profiteering, and a permanent wealth tax on the richest 1 per cent, at rates high enough to significantly reduce the numbers and wealth of the richest people and redistribute these resources. Gabriela Bucher, Executive Director of Oxfam International, summed up the potential benefits: 'Taxing the super-rich is the strategic precondition to reducing inequality and resuscitating democracy. We need to do this for innovation. For stronger public services. For happier and healthier societies. And to tackle the climate crisis, by investing in the solutions that counter the insane emissions of the very richest.'[38]

It is surely no coincidence that Finland was ranked the happiest country in the world for the sixth year running in 2023, demonstrating once again that happiness does not appear to come from having more than everyone else, but from feeling part of a more equal society with sufficient social support, access to good education, freedom to make decisions and low levels of corruption.[39] All this, in an economy where paying progressive taxes is regarded by the majority as a worthwhile measure for a good life.[40] It appears to be an idea whose time has come. Three quarters of British people now support a wealth tax.[41]

* * *

Diary: 23 May 2023, London

I am sitting inside Shell's AGM at the ExCeL centre in London. Dirk Campbell, whom I last saw on television politely interrupting Jacob Rees-Mogg to enlighten the National Conservative conference as to the characteristics of fascism, is standing on a chair telling the Board and assembled shareholders (which include me): 'Your record-breaking profits are producing record-breaking floods, famines and heatwaves. We are putting you on notice that either you shut down Shell or we will do it for you.' Then a choir in the row directly in front of me leaps to their feet singing 'Go to hell Shell, and don't you come back No more, No more, No more, No More!' to the tune of 'Hit the Road Jack'.[42] Chair of the Board, Sir Andrew Mackenzie, asks them to sit down and have 'a civil debate about this later' by waiting for the Q and A, but as they persist, he calls security to take them out. Some are carried, some walk. And as Sir Andrew repeatedly tries to start the meeting, speaker after speaker gets up to interrupt. The man on one side of me is telling me these are 'lunatics who need to go to the asylum'; the woman on the other is writing a speech about her Nigerian grandchildren, accusing Shell of poisoning millions in the Niger Delta. Two activists are blocked by a human shield of security guards as they try to rush the stage. More speakers get up and are carried out.

Shell's profits of £32.2 billion in 2022 were the biggest in its history, and yet it's not using these to fund a rapid clean energy transition. Instead it gave its outgoing CEO £9.7 million in pay and bonuses last year, a 53 per cent pay rise on the year before.[43] After further record profits in the first three months of 2023, it has 'showered shareholders with more than $6bn'.[44] Meanwhile, its annual report acknowledges that climate activists' challenges and protests over new and existing permits could 'delay or prohibit operations in certain cases'; that some investors were choosing to divest; stakeholder groups were pressuring banks to stop financing fossil fuel projects; and 'some financial institutions have started to limit their exposure to fossil fuel projects. Accordingly our ability to use financing for these types of future projects may be adversely affected.'[45] It sounded like pressure was working. So when Fossil Free London, a loose network using creative direct actions to target fossil fuel companies and their backers, asked for people to buy shares in Shell and try to close the AGM down, I decided to join them. I wanted to look in eye the people whose combined pay could fund 390 nurses.[46] The meeting is not shut down, but it is delayed for more than hour. Moreover, having offered a 'debate', Sir Andrew has no choice but to listen to the barrage of climate-related points that follow in the Q and A.

It feels as if everyone is here: the proposers of the 'Follow This' motion to reduce emissions by 2030; representatives of the Church of England Pensions Board voting to get rid of the current board of Shell; the litigants from the Milieudefensie case against Shell in Holland, pointing out Shell's complete failure to meet that 45 per cent reduction target directed by the court. Sir Andrew and new CEO Wael Sawan's answers are predictably avoidant and depressing. Apparently if they do not pump oil someone else will do so. When I ask how they plan to reduce fossil fuel air pollution causing the deaths of 8 million people a year, they express concern but have no plan. Nor do they respond to my reminder that António Guterres says they have 'peddled a big lie' and 'fossil fuels are a dead end'. Caroline Dennett, the safety officer who resigned from Shell in 2022, speaks last, cataloguing in detail Shell's disastrous performance in taking care of the safety and wellbeing of its workers and the environment in the Niger Delta. She asks investors if they want their hands stained from dabbling in blood money, or whether it is 'time to hold Shell to account and use your power to clean them up or disengage entirely'.[47]

* * *

The 'Follow This' motion was rejected by 80 per cent of investors, but actions both inside and outside the meeting got detailed global coverage. It was just one of a series of disrupted AGMs that took place at that time. On 26 May, as French police teargassed nonviolent protesters blockading the TotalEnergies AGM in Paris, shareholder support inside for their 'Follow This' motion doubled from previous years to 30 per cent.[48]

The next day, Dr Patrick Hart jumped over the barrier onto the rugby pitch at Twickenham and ran across it into the middle of the Premiership final. I watched the film on Twitter as he and fellow JSO activist, construction worker Samuel Johnson, threw orange paint powder into the air, as the crowd booed, threw beer cans, and they were caught and escorted off. The next day both were charged with aggravated trespass and criminal damage. Patrick said he acted 'because it's my duty as a doctor. The climate crisis is the greatest health crisis humanity has ever faced. People are dying now and more will die every day unless we stop new oil, gas and coal.'[49] The game was stopped for five minutes, no one was hurt or prevented from doing essential work, and the orange-dyed cornflour used was easily washed away.

On that same day, Labour leader Keir Starmer confirmed the commitment he made during the World Economic Forum Annual Meeting at Davos in January 2023. A future Labour government would block all new oil and gas developments in the North Sea.[50] Are all these events completely unrelated?

The arguments about what kind of protest works have been raging for the entire period that I have been writing this book. And however vitriolic, the almost daily discussion about the use of disruptive nonviolence in a democracy is itself a sign of hope, because it suggests that civil disobedience has legitimacy. From my own experience of being castigated while living at Greenham, unpopularity is not a measure of effectiveness. Many nonviolent protest movements have been unpopular at the time of their actions, regardless of the specific choice of tactics. The majority of Americans did not approve of Freedom Riders, bus boycotters or those doing sit-ins at the time their actions took place. In 1966, a year after the Voting Rights Act was passed, 85 per cent of white and 30 per cent of Black people nationally, believed that civil rights demonstrations by Black people actually hurt the advancement of civil rights.[51]

The Social Change Lab, a not-for-profit research team looking at social movements, has reviewed literature on protest and found that overall nonviolent protest movements in North American and European democracies could be effective in achieving their aims around the issues of 'civil rights, climate change, and social welfare'. In particular, there was an impact on 'public opinion, public discourse and voting behaviour'. Context mattered: movements were more successful if they highlighted issues where the public were already supportive, and where they had been recently covered by the media.[52]

As to the specific effect of nonviolent disruption, Herbert Haines, who coined the term 'radical flank effect', showed in a 1984 paper that it could be positive. He discovered that 'the activities of relatively radical black organizations, along with the urban riots, stimulated increased financial support by white groups of more moderate black organizations, especially during the late 1960s'.[53]

The Social Change Lab researchers' review found something similar. Overall the public were not put off. On the contrary, 'a movement with a nonviolent radical flank is likely to be more successful than a movement with no nonviolent radical flank'. The suggested explanation is that 'a positive radical flank leads to people being more likely to identify with a more moderate group [...] without undermining (or increasing) support for broader movement issues'.[54] This means that the more disruptive nonviolent tactics don't actually change people's views about an issue like the climate crisis either way. But if someone is already concerned, these tactics, even if they don't like them, may mobilise them to take some kind of more moderate action.

This is exactly what Social Change Lab found when they did specific research on the impact of Just Stop Oil's M25 blockades in November 2022. Using the market research company YouGov, they surveyed a nationally representative sample of 1,415 people, both before and after the JSO week-long campaign, 'to understand whether a Just Stop Oil campaign had any impact on support for or identification with more moderate climate organisations'. They found that 'increased awareness of Just Stop Oil after their M25 protest campaign was linked with stronger identification with and support for Friends of the Earth'.[55]

Certainly, while I have friends and colleagues discomfited by disruption, I have yet to have someone tell me 'I hate your tactics so I will no longer believe all this rubbish about the climate crisis.' On the contrary, if the topic comes up it is often because the media have covered someone like Patrick, and they are looking for something else they can do. Like the lady I met on her way into Barclays Bank in Penzance when my Dolphin friends were doing street theatre outside. She watched for a while, and then turned to me—leafleting as usual—and said she was worried, 'but I don't like that sitting in the road'. When I suggested she could go in, pull her bank account and tell them why, and gave her a leaflet explaining how to do it, she was delighted.

Many of my health worker colleagues support slow walking, and some have joined the actions. I admire the sacrifice and courage required in risking arrest, dealing with an angry public and on occasion direct physical aggression. It was good to see Lord Deben acknowledge in a recorded video on Twitter, that while he did not want 'to interfere with what the law has done', he recognised that 'these people are doing what they are doing because we are not moving fast enough, and we have to'. He wanted everyone to recognise the urgency: 'The whole political establishment should realise that these protests come from people who are desperate because they know their future is being imperilled, because we are not doing enough. [...] They ought to make themselves a nuisance in every circumstance they can, because we have to act now.'[56]

This was just prior to the Climate Change Committee publishing a withering annual report at the end of June 2023 on the government's failure to meet its own climate change targets. Lord Deben, as outgoing chair, castigated the government for its lack of leadership in a 'global emergency' and said that supporting new coal mines in Cumbria and oil and gas fields in the North Sea was 'utterly unacceptable'.[57]

However, a medical colleague wrote to me about a recent slow walk in which she was involved where she faced a lot of very angry traffic and abuse, as well as risk from cars cutting in, accelerating fast past her and driving over the central reservation. She described it as both 'stressful and unsafe'. When two distressed teenagers got out of their cars and begged them to end it because they had to get to their GCSE exams, they stopped the march: 'Overall I'm not sure how effective they are. I get the reasoning and the government leaves us very little option for protest [but] even sporting events create too much negativity and hostility if play is interrupted.' She wanted to focus on actions that were 'non-disruptive to the general public, such as museums which are sponsored by oil companies, banks who fund fossil fuel extraction and government offices'.

I agree. Personally, I have become increasingly ambivalent about any form of general, prolonged road obstruction as a tactic. This is not because these actions are too radical, but rather because of the risk of harm as described above. There is no way to assess who is in that obstructed queue. And as health professionals, we have a particular responsibility to 'do no harm', as a 2020 *Lancet* editorial endorsing civil disobedience made clear.[58] Also, long lines of slowed or idling cars increase local air pollution, something we are trying to stop. Like my colleague, I think we have to focus on creative and communicative nonviolent disruption, aimed at those directly responsible for the damage, or those who support it through financial or cultural ties. Fossil Free London's action at the Shell AGM was a good example.

But no one has left Health for XR because some doctors support slow walking, or because Patrick Hart jumped onto a rugby pitch. On the contrary, the network has continued to grow. GP Chris Newman argues that even though disruptive tactics make him feel uncomfortable, they do keep the issue in the news, build pressure, and create a sense of urgency that pulls other people along. So 'the mainstream can step in behind us to take advantage of the shifting Overton window'. For example, health colleagues already developing greener practices could push hospitals to shift to plant-predominant diets, 'grow food in hospital grounds in place of ever-expanding car parks, and ensure colleagues and staff understand not just what we need to do, but the reasons why it is of the utmost urgency'.[59]

On the other hand, Social Change Lab's review suggests that 'violent radical flanks (involving violence to humans, rioting and some forms of property destruction)', appear to have a negative effect on the overall movement.[60] This would suggest that Andreas Malm's call for more 'sabotage' of essential infrastructure, combined with, where necessary, violent confrontation with the State, would be the wrong way to go. His argument that this combination is necessary because those in power will never make concessions, both fails to protect others from harm, and ignores the fact that those in power have access to far more violence than we do.[61] Nonviolent action, as Gene Sharp, Michael Randle and Erica Chenoweth make clear in all their writings, is a much more effective means of undermining them.

Indigo Rumbelow, a spokesperson for JSO, made a similar point to me at a meeting in London: 'We are not engaged in a war. If we are the State is stronger, it has helicopters, dogs, police, we could not win. We are here to change the story—how people understand oil—that it's toxic to democracy and it's killing people. For that we need a media presence. The capital is the heart of power.'[62] We were chatting in Friends Meeting House near Trafalgar Square, during the April 2023 action. Wildlife expert and campaigner Chris Packham had just given a talk encouraging people to support JSO, as well as pointing out that they were part of 'an ecosystem filled with different forms of activism'.

What gives me the most hope is that I see that ecosystem emerging. Indeed, there is so much going on it is hard to keep up.

At the end of March 2023, pushed by youth climate activists in the Pacific island nations, the UN General Assembly adopted a resolution calling for the International Court of Justice to give an advisory opinion as to 'what are the obligations of States under international law to ensure the protection of the climate system and other parts of the environment from anthropogenic emissions of greenhouse gases for States and for present and future generations', and what the legal consequences of failing to live up to these obligations would be. Britain was one of 120 co-sponsors. The advisory opinion would be just that and not legally binding, but it would be influential, creating an atmosphere in which countries could be held accountable and climate litigation easier to pursue.[63] In April 2023, Greenpeace and Uplift were granted permission to challenge the UK government decision to issue new oil and gas licences through a judicial review.[64]

The court's heavy-handed response to climate protesters is also being challenged. In March 2023, Judge Silas Reid extended his reach outside the Inner London Crown Court to the street, where 68-year-old former social worker Trudi Warner was holding up a placard reading: 'Jurors: You have an absolute right to acquit a defendant according to your conscience.' Inside the court, four Insulate Britain activists were on trial for causing a public nuisance. Reid had Warner arrested for contempt, and her case was then referred to the Attorney General Victoria Prentis KC MP.[65] A month later, during another Insulate Britain trial, twenty-four activists, including lawyers, health workers, teachers, faith leaders and a former policeman, protested over two consecutive days by sitting quietly outside on the street holding up similar signs. They sent Judge Reid a letter of explanation protesting his continuing ban on people explaining their motivations in court, as well as Warner's arrest:

> There is a right to freedom of expression in this country, which may only be restricted as prescribed by law and as necessary and proportionate in a democratic society. There is no law against upholding the law. It can never be necessary and proportionate to prevent a peaceful demonstration against manifest injustice.[66]

No one was arrested or asked to move at the time, but Judge Reid referred these protesters to the Attorney General. According to a press release issued by Plan B in June, the Attorney General had to choose between 'giving the green light to informing the public of a jury's right to make a decision on their conscience, whatever the judge directs them to do', thus undermining judges who have been banning protesters from speaking in court; or commencing a prosecution that would inadvertently shine a spotlight 'on the highly contentious issue of judges seeking to suppress one side of the story in jury trials'.[67] Meanwhile, charity director Mary Adams, one of those on trial, explained to the jury that she had blocked the road in 2021 because she cared about public good and future generations. She knew that she risked Reid charging her with contempt, but 'when context and motive are stripped away, where is the justice in the law? The climate crisis is a fact, it's not a "personal belief" as described in this court.' This time Reid neither cleared the court, nor charged her with contempt, and the testimony appears to have had an impact. The nine-person jury could not reach a verdict and was dismissed on 18 May. At the time of writing the decision on a retrial is yet to be made.[68]

On the other side of the world, *El Periódico* editor José Rubén Zamora was sentenced to six years in prison in June 2023.[69] However, all charges against Bernardo Caal Xol have been dismissed, and he continues to campaign to protect water and land, and for others similarly criminalised.[70]

Acts of non-cooperation do bring results. Following the uproar caused by the Declaration from Lawyers Are Responsible that they would avoid representing fossil fuel clients, Stephen Kenny KC, chair of the Bar Council's Ethics Committee, clarified that lawyers do not always have to follow the cab rank rule in taking cases. At a Bar Council debate on the issue in April 2023, Kenny explained that regulations allowed a barrister to refuse instructions, 'if there is a real prospect you are not going to be able to maintain your independence. So, if you are genuinely afflicted by conscience, such that you cannot properly do your job as an advocate [...] you do not have to act.' This came after the Law Society issued crystal clear guidance to all solicitors: 'Solicitors may also choose to decline to advise on matters that are incompatible with the 1.5°C goal, or for clients actively working against that goal if it conflicts with your values or your firm's stated objectives.'[71]

And divestment continues. In June 2023, the Church of England announced that its multi-billion pound endowment and pension funds would be divested from fossil fuel companies, including Shell, because of their backtracking on pledges to cut production. As the Archbishop of Canterbury, Justin Welby, explained, they were failing to help the church 'achieve the just transition to the low carbon economy we need'.[72] The British Museum also decided to end its twenty-seven-year relationship with BP.[73] Meanwhile, insurers and banks have continued to pull their support from the East African Crude Oil Pipeline.[74] This is in response to a transnational campaign involving a whole range of actions, including thousands of emails from across the world to company staff asking them to talk to their bosses; actions outside company facilities; shareholder pressure; and continuing extraordinary courage from activists who face arrest and jail simply for speaking out and protesting the pipeline.[75] Extinction Rebellion is just one of a global network of climate and ecological justice and finance groups from both the Global South and North, working together.[76]

Local action also has an effect. In February 2023, Derek Thomas, our Conservative MP for St Ives, endorsed the Climate and Ecology Bill, going through parliament as I write.[77] This Private Members' bill calls for the rapid cutting of emissions to keep temperatures below 1.5°C; prioritising nature and reversing its decline by 2030; taking responsibility for our global ecological footprint; and ensuring a fair transition. Significantly, it calls for a Climate and Nature Assembly to help government and parliament develop the strategy to achieve these goals.[78] The endorsement was surprising given Derek Thomas's consistent record of voting against measures that might address the climate crisis.[79] It shows that the long slog of grassroots campaigning and sustained lobbying from a wide coalition, including farmers, schools, faith groups, the Women's Institute, as well as environmental campaigners, can produce results.

Our Tory-led county council also changed its mind, reversing its rejection of the same Bill in 2021 and giving it unanimous support in April 2023, as well as promising to lobby government themselves. Possibly, those regular People's Assemblies that the council had initially tried to evict contributed. Disruption can lead to dialogue and change.[80]

Increasingly, local activist groups come out in support of each other, whether it is climate activists protesting the eviction of refugees at a local hotel, supporting strikers on picket lines, or housing campaigners turning up to protest at dirty water or to help in beach cleans. When Dolphins organised two meetings on community energy, bringing together the experience and knowledge of experts across the county, they were our best-attended meetings to date.

The philosopher Rupert Read would be pleased. In a March 2023 video on YouTube he argued that a larger moderate flank, what he calls a 'climate majority' could be created through people's direct personal and professional connections, through actual lived relationships. He called for people to build up community resilience by working together on joint community projects.[81] This has been happening in Penzance for some time, and it is one of the first towns to become plastic free. Sustainable Penzance now works on building community resilience across multiple areas including transport, food, business and energy.[82] Along with groups as diverse as the Women's Institute and Extinction Rebellion, it participates in the Town Council's climate emergency sub-committee, which has pushed local initiatives such as a traffic-free town centre and divestment.[83]

Far from fragmenting, as some in the media suggest, what I have sensed from every direction is a growing fluidity and flexibility within the climate and environmental movements; increasing respect for different tactics; a willingness to learn and adapt quickly according to context; and mutual support. Personally,

Asmamaw and I don't see ourselves as 'belonging' to one particular group. We work with any we see as taking effective action to bring about climate and ecological protection and global justice. This includes lecturing to educate others, petitioning, lobbying, joining specific nonviolent direct actions, supporting others engaged in civil resistance and continuing relief work.

I asked Roger Hallam about the necessity of using a broad range of nonviolent tactics at the Falmouth meeting. He agreed, telling us that different kinds of non-cooperation, including tax refusal and strikes, were 'an essential part of a broader strategic civil resistance strategy against the whole system itself'. He was excited by 'the increasing awareness and willingness of people to engage in cooperation, across cultural and political boundaries'. His current focus was on how to bring movements together with multiple strategies and tactics but a 'certain strategic coherence [...]. Just Stop Oil is one part of a wider fabric that's developing.'

We need that wider fabric. Chenoweth argues that 'movements that grow in size and diversity are more likely to succeed', and that 'movements that do not solely rely on protests, demonstrations, and digital activism, but also build power through parallel institutions, community organizing, and non-cooperation techniques are more likely to build an effective and sustainable following'.[84] Chris Newman gave me an image: 'Some say the movement is splintering into many pieces, but I see it more like a tree spreading different branches to catch maximum sunlight. And each branch has to remember, we all have the same roots.'[85]

What can't be denied by even the most sceptical, is that three quarters of people in Britain are worried about climate change (second only to anxiety about the cost of living); three quarters think it is going to get hotter in Britain in the next six years; and three quarters of the population have tried to make some changes to address it.[86] The evidence shows that the majority of people are worried and want to do something. What matters now is creating that biodiverse landscape of action that makes it possible for all of us to act effectively, with mutual respect and with the urgency needed to create a genuinely just and planet-friendly society. So I will end this book where I began, with Greta Thunberg, who in *The Climate Book*, sums up what she believes that transformation requires:

> Individual action and systemic change to work hand in hand [...] We need both. Any suggestion that we can have one without the other—or that any single solution or idea is more important than all the others—is pretty much guaranteed to be aimed at slowing us down. [...] We as individuals should use our voices, and whatever platforms we have, to become activists and communicate the urgency of the situation to those around us. We should all become active citizens and hold the people in power accountable for their actions, and their inaction.[87]

'Drops of water break the rock', Asmamaw reminds me. This is an emergency. We need waterfalls.

EPILOGUE

DIARY

19 OCTOBER 2023, LONDON

I am outside the Intercontinental Hotel near Hyde Park. Inside, at the Oil and Money conference—or rather the 'Energy Intelligence Forum 2023', as they renamed it this year—'energy leaders' are taking part in 'high-level networking', debating and shaping 'sustainable solutions to the energy challenges of the 21st century'.[1] I wish they were. But panel discussions on 'how best to navigate climate, supply security and shareholder returns' suggest the real agenda.

'The elites of the oil and money conference, they have no intention of transition. Their plan is to continue this destructive search for profits. That is why we have to take direct action to stop this and to kick oil money out of politics', Greta Thunberg said in her speech to some hundreds of activists from Fossil Free London blocking both ends of Hamilton Place two days ago. Greta, along with five others, was arrested for obstruction. But Shell CEO Wael Sawan had to give his speech by video and the CEO of TotalEnergies and a few other executives had to cancel their contributions altogether.[2]

Today, as the rain pours down, friends and colleagues from Health for XR are holding an inquest. Fourteen of them are lying down as shrouded corpses, spread across half the road. Others give testimony to the 'coroner' as to how each one died: a 16-year-old dying of cholera in Malawi because flooding contaminated her water supply with sew-

age; an elderly woman dying of heatstroke in Brazil; a farmer committing suicide after his crops failed because of drought in Kenya; a woman dying in childbirth in Pakistan because she could not get medical help when the bridges were washed away … not statistics, real people, individuals dying now. The police do not intervene. Workmen on a scaffolded building opposite the hotel, who were shouting when we began, have now become silent. Some have come down to watch and are filming.

Why am I standing here? Because in the last three months I have felt as if I exist in two parallel worlds. In one, the planetary emergency escalates: our oceans are warmer than they have ever been;[3] Antarctic Sea ice suffered its greatest depletion this year.[4] Those predicted heatwaves have occurred: we have just experienced the hottest July, August and September ever recorded.[5] We are in 'the era of Global boiling', as António Guterres put it.[6] Wildfires drove 20,000 tourists off the island of Rhodes and burned an area the size of Greece in Canada, sending smoke pollution as far as New York and Florida.[7] A devastating monsoon season across Asia killed hundreds and displaced thousands. And in Libya, heavy rains, made 50 times more likely and up to 50 per cent more intense by human-induced climate change, burst two dams and inundated the city of Derna, sweeping thousands into the sea.[8] The Guardian *estimates that in the three months since I handed in what I hoped was a completed manuscript to my publisher, from June to early September 2023, 'extreme weather disasters took more than 18,000 lives, drove at least 150,000 people from their homes, affected hundreds of millions of others and caused billions of dollars of damage'.[9]*

But in that other world this is just weather. This is the 'new normal' in which Italy's tourism minister Daniela Santanchè can tell people that 'the high temperatures are physiological in this season and do not compromise in any way our tourist offer'.[10] Does she know that six of the Stockholm Resilience Centre's nine planetary boundaries have now been crossed? To put it simply, as the report by the

EPILOGUE

Centre explains, our overuse of land and freshwater, our destruction of nature, our use of plastics and pesticides, as well as our changes to the climate, have moved us out of the 'safe operating space' in which humanity developed over millennia. If the earth were a human body and the planetary boundaries were blood pressure, it would be too high, Katherine Richardson, one of the authors, stated.[11] *It is a good metaphor. Yet still it does not cut through to those with the power to make the changes necessary.*

On the contrary, as the evidence of planetary dysfunction continues to pile up by the day, the British government, alongside others across Europe, continues to close the space for nonviolent protest and to vilify and criminalise those sounding the alarm. This is despite Michel Forst, Aarhus Convention Special Rapporteur on Environmental Defenders, pointing out that such 'vilification campaigns are a threat to democracy'; and calls for more dialogue and less repression from Dunja Mijatović, the Council of Europe Commissioner for Human Rights.[12] *Morgan Trowland and Marcus Decker, in jail for suspending themselves over the Dartford Crossing, had their appeals against their lengthy sentences turned down, even though Daniel Friedman KC, representing the pair, argued that these were 'the longest ever handed down in a case of non-violent protest in this country in modern times' and a 'disproportionate interference' with the activists' rights to free speech and protest. Lady Justice Carr disagreed. The sentences were a 'legitimate' deterrent, she claimed, because Trowland and Decker were repeat offenders and the protest, although nonviolent, 'had extreme consequences for many, many members of the public.' Permission to appeal to the Supreme Court was denied.*[13]

It is not just criminalisation. Environmental protesters are increasingly faced with civil injunctions that can result in up to two years in jail and unlimited fines if breached. But contesting a breach and losing can result in even larger fines. National Highways and Transport for London are charging several named Insulate Britain and Just Stop Oil activists for the legal costs of obtaining the injunction, even if not breached.[14]

Other restrictions loom. The government hopes to ban UK public bodies 'from imposing their own boycott or divestment campaigns against foreign countries and territories.'[15] *Campaigners would never have succeeded in helping to end Apartheid in South Africa if such a law had been in place at that time. More than 60 civil society organisations are protesting, pointing out that this law, if passed, 'presents a threat to freedom of expression, and the ability of public bodies and democratic institutions to spend, invest and trade ethically in line with international law and human rights.'*[16] *Meanwhile, Trudi Warner has been told that she will be prosecuted for contempt of court for holding up that sign in March 2023. Some of those who have held up signs in her support are being investigated for 'perverting the course of justice', an offence which carries a maximum sentence of life imprisonment.*[17]

So how does one cut through? How does one reach a prime minister who chose the same month those floods in Libya drowned at least 4,000 people and left 9,000 missing to announce his break-up with his party's already inadequate net zero commitments, and his new love affair with motorists. His gifts to them include the removal of tiresome restrictions on speed limits and a longer opportunity to buy new fossil fuel–fed cars.[18] *And no worries about running out of those fuels. This July, Rishi Sunak promised 100 new drilling licences for oil and gas, and, at the end of September, he gave the Norwegian company Equinor permission to exploit Rosebank, the largest untapped oil and gas field off the Shetland Islands.*[19] *Never mind that the UN just made clear that avoiding climate disaster requires 'phasing out all unabated fossil fuels'.*[20]

That U-turn on net zero was condemned not just by environmentalists, but business leaders, car manufacturers and fellow Tory party members.[21] *But Sunak was not listening, nor was he going to the special Climate Ambitions Summit at the UN General Assembly. So, he missed hearing Guterres point out that 'humanity has opened the gates of hell', while calling again for countries to take on the 'entrenched interests raking in billions from fossil fuels'.*[22]

At our inquest, Jamie sounds the last blast on his trombone. There is a minute's silence, then a group of us sing the 'Dies Irae' from Mozart's Requiem: 'Day of wrath, day of anger will dissolve the world in ashes'. I am angry. I can smell the vehicle fumes in the air. Last month, we learnt that 98 per cent of Europeans live in areas with toxic levels of air pollution.[23] Sunak is right: 'politics has failed.' I would love to have a government that really does have the 'courage to change direction', as the prime minister claimed in his speech to the Conservative Party Conference at the beginning of October. The biggest 'public health intervention in a generation' could be seriously addressing the climate and ecological crises by taking on those 'vested interests standing in the way of change',[24] namely the fossil fuel industry, some of whom are schmoozing inside the Intercontinental right now.

But the main reason I am here is because I am hopeful. According to an August survey by YouGov, 82 per cent of the British public now think climate change is very, or fairly, important.[25] Another more recent poll suggests only 22 per cent trust Sunak to tackle the issue.[26] His counterparts are moving in a different direction. The Welsh government has brought in 20mph speed limits to make the streets safer and healthier. Scottish first minister Humza Yousaf has said that granting new oil and gas licences, including Rosebank, is 'climate denial'.[27]

Repressive legislation and media vilification does not seem to be working. That YouGov poll showed that while 68 per cent disliked Just Stop Oil protesters, the majority thought nonviolent disruptive protesters should not be imprisoned, despite the constant demonisation. I watch on social media as American tennis champion Coco Gauff says climate change is real and she 'wasn't pissed' at the activists who disrupted her US Open semi-final match. 'Moments like this are history defining moments. [...] I always speak about preaching what you believe in. It was done in a peaceful way'.[28] When two climate protesters from 'Renovate Switzerland' interrupted Bruckner's Symphony

No. 4 at a music festival in Lucerne, the conductor, Vladimir Jurowski, asked the audience to let them speak. When one of the protesters was heckled, Jurowski said he would leave the stage unless she was allowed to finish. She did. No one was arrested. The video has had millions of views.[29]

One action leads to another. Early in the morning on 25 September, 32 friends and colleagues sat down in silence outside Truro Crown Court holding signs saying 'Jurors have an absolute right to acquit a defendant according to their conscience'. They were just some of the more than 250 people holding signs outside Crown Courts around the country that day, in response to a call from a new group: Defend Our Juries. There was no obstruction. It was not an act of civil disobedience but a dignified restatement of rights that the government and some judges prefer to keep hidden.[30]

Saying no is gaining traction. Two senior executives resign from Shell in response to CEO Wael Sawan's backtracking on the planned transition to clean energy.[31] *Celebrities reject their nominations by the British LGBT Awards and judges withdraw when they learn of its sponsorship by Shell and BP.*[32] *Greta Thunberg pulls out of the Edinburgh Book Festival, and climate activist and author Mikaela Loach stages a walkout with her audience, because festival sponsor Baillie Gifford invests in fossil fuels.*[33] *Former Australian rugby captain David Pocock says oil companies should not sponsor World Cups, criticising TotalEnergies for trying to 'greenwash' its image.*[34] *And in a sign of mainstream support, Which?, the popular consumer choice organisation, website and magazine, has put six banks, including Barclays and JPMorgan Chase, on their 'Red List'. The deputy editor advises readers to switch to one of the three identified eco providers 'if they are uncomfortable with their money being invested in the fossil fuel industry and other projects which could be damaging to the environment.'*[35]

There is good news on the other side of the world. In Ecuador, citizens voted decisively in national referendums to stop oil drilling in the

Yasuní National Park and gold mining in the highlands near Quito. In both cases, public education and legal campaigns by environmental activists played a significant role. This is just part of a growing struggle to block oil and gas extraction across the Amazon.[36]

In Guatemala, Semilla's anti-corruption candidate, Bernardo Arévalo won the August presidential election with 58 per cent of the vote. His victory vindicated the process of quiet grassroots organising and building coalitions of social movements that I witnessed last year. And as the current attorney general tries to undermine this result, including suspending Semilla as a political party, a nationwide campaign of nonviolent protest has sprung up in response. As I stand here in London, indigenous groups, students, workers and communities are in their second week of demonstrating and blocking roads across Guatemala. Bernardo Caal Xol posts pictures on Facebook of packed streets, some filled with dancing crowds. One catches my eye: four women in a mass demonstration blocking a road hold a blue banner with a Spanish slogan that translates as 'I am sorry that my protest collapses the traffic, but your indifference is collapsing the country'.[37]

Meanwhile in the Netherlands, a 27-day blockade of the main road into The Hague by thousands of Dutch Extinction Rebellion activists to protest their government's fossil fuel subsidies has pushed the lower house of the Dutch parliament to ask the cabinet to come up with a phase-out path for those subsidies. The powerful image of an orchestra playing Mozart's Requiem in the road gained international attention, while the daily use of water cannon and the more than 9,000 arrests did not stop the protesters. Rather, it increased support. More than 70 per cent of the population now favour the abolition of subsidies. 'Civil disobedience works', XR Nederlands says on its website.[38]

So does insisting that governments uphold the law. Naturalist Chris Packham argues that our 'reckless and irresponsible' prime minister is 'acting illegally' in changing the timeline for net zero, and contravening the 2008 Climate Change Act. He has demanded a judicial review.[39] *The advocacy group Liberty are taking Home Secretary Suella Braverman to court for extending anti-protest legislation without consulting parliament.*[40]

In Montana, Judge Kathy Seeley agreed with 16 young people that youth in the state have a 'fundamental constitutional right to a clean and healthful environment, which includes climate'. This means that state agencies must now consider the effects of fossil fuel projects, a judgement that may have significance beyond that state.[41] *Perhaps it will encourage those six Portuguese young people who are suing 32 European governments for failing to protect them from climate catastrophe. They had their day in front of 22 judges at the Grand Chamber of the European Court for Human Rights in Strasbourg at the end of September. The 86 lawyers representing the European states argued that the Court has no jurisdiction over matters that should be resolved at a national level, and made the astonishing claim that the climate crisis posed no direct threat to the children's lives or health.*[42] *Where were they this summer?*

In the afternoon, we repeat our inquest again, outside the Embankment offices of JPMorgan. Through the glass wall I watch a video playing on a large screen in the office lobby, displaying snippets of the JPMorgan story. It tells us that the bank stands for 'honesty, integrity and sound judgement'. This from the company that has invested US$434 billion dollars in fossil fuels since the Paris agreement,[43] *even though their own economists warned clients in 2020 that if business as usual continued they could not 'rule out catastrophic outcomes where human life as we know it is threatened'.*[44] *'The problem lies with our imagination as much as our cognition', Lancet editor Richard Horton suggests. 'We still cannot imagine a time when our species could contemplate extinction.'*[45]

Outside the office the white shrouded corpses fill the pedestrian space. We sing the 'Dies Irae' again. I find watching the corpse tenders placing a rose on each body a second time even more upsetting than the first. A crowd of people have stopped to watch and listen.

There is no barracking or mockery. As in the morning, the police do not intervene. The coroner asks everyone to imagine what a different future might look like, one 'grounded by connection, care and collaboration', and 'people realising, as if for the first time, that together, they might have the power to change the way ahead'. Perhaps for one moment we have helped them to do just that.

I was teaching on 'grief and loss in conflict and disasters' last week, as part of an online annual master's course on Mental Health in Complex Emergencies. Next week, I will teach on 'stress-related disorders'. The students are humanitarian professionals from across the world, including Afghanistan, Bangladesh, Libya and the Middle East. The question that always comes up in our Zoom classes is the one that has dogged my life since I first went to live at Greenham. Where does our duty lie? Is it to mitigate the grief and stress caused by the miseries of these times? Or is it to try to prevent the escalating numbers of conflicts, both visible and invisible; the unnatural disasters; the poverty, hunger and dislocation that are causing that stress and grief in the first place? Strangely, I find myself in agreement with disgraced former health secretary Matt Hancock who, in giving evidence to the UK Covid-19 Inquiry, said the government got things 'completely wrong' in pandemic planning, because it focussed on preparing for 'the consequences of a disaster' instead of 'how do you stop the disaster from happening in the first place?'[46]

My answer to the students is that we need to do both. We must reach out to all those affected with empathy and whatever professional skills we have, to mitigate individual suffering. But we must also take every action we can to try to prevent the horrors that underpin it. And we must imagine that healthier, happier, more equitable future, so that we can create it. We cannot do it alone, but it can be done. That is why I am standing here.

NOTES

INTRODUCTION

1. NPR, 'Transcript: Greta Thunberg's Speech At The U.N. Climate Action Summit', 23 September 2019, https://www.npr.org/2019/09/23/763452 863/transcript-greta-thunbergs-speech-at-the-u-n-climate-action-summit (last accessed 9 November 2023).
2. Concern Worldwide, 'Refugee, Migrant, IDP: What's the difference?', 4 May 2022, https://www.concern.org.uk/news/refugee-migrant-idp-whats-difference (last accessed 9 November 2023).
3. Science and Security Board, 'A time of unprecedented danger: It is 90 seconds to midnight, 2023 Doomsday Clock Statement', Bulletin of the Atomic Scientists, 24 January 2023, https://thebulletin.org/doomsday-clock/current-time/ (last accessed 9 November 2023).

1. ACCIDENTS AND EMERGENCIES

1. E. P. Thompson, *Protest and Survive*, (Nottingham: Russell Press for the Campaign for Nuclear Disarmament and the Bertrand Russell Peace Foundation), p. 2.
2. Ibid., p. 21.
3. Ibid., pp. 15–16.
4. This was not factually correct. In his State of the Union address Carter condemned the action, withdrew the US ambassador, asked the Senate to delay consideration of the SALT II arms limitation agreement and imposed cultural and economic sanctions. Jimmy Carter, 'Address to the Nation on the Soviet Invasion of Afghanistan', 4 January 1980, The

American Presidency Project, https://www.presidency.ucsb.edu/documents/address-the-nation-the-soviet-invasion-afghanistan　(last accessed 14 October 2023).

5. From the speech by Lord Mountbatten on 11 May 1979, published in *Pugwash Newsletter* 17, no. 4 (April 1980), p. 20.

6. Thompson, *Protest and Survive*, p. 22.

7. 'The Appeal for European Nuclear Disarmament', Bertrand Russell Peace Foundation, April 1980, http://www.russfound.org/END/EuropeanNuclearDisarmament.html (last accessed 1 June 2023).

8. Stan Openshaw, Philip Steadman and Owen Greene, *Doomsday: Britain after Nuclear Attack* (Oxford: Blackwell, 1983).

9. Lynne Jones, Letter to the editor: 'Civil defence', *The Lancet* 316 (1 November 1980), pp. 976–7.

10. J. H. Humphry et al., Letter: 'The consequences of nuclear war', *British Medical Journal* 281 (29 November 1980), p. 1497.

11. Lynne Jones, 'On common ground: The women's peace camp at Greenham Common', in L. Jones (ed.), *Keeping the Peace* (London: Women's Press, 1983), pp. 79–97.

12. Ibid., pp. 79–97.

13. '2018–2019 Mozambique humanitarian response plan revised following cyclones Idai and Kenneth, May 2019 (November 2018–June 2019)', UNOCHA, 25 May 2019, https://reliefweb.int/report/mozambique/2018-2019-mozambique-humanitarian-response-plan-revised-following-cyclones-idai (last accessed 1 June 2023).

14. Tsvangirayi Mukwazhi 'On cyclone-shattered island in Mozambique, shock and debris', Associated Press, 2 May 2019, https://apnews.com/article/health-cholera-ap-top-news-africa-international-news-42f8975970fd4242a412c7d2e602486c (last accessed 1 June 2023).

15. Francesco Basseti, 'Environmental migrants: Up to 1 billion by 2050', Foresight, 22 May 2019, https://www.climateforesight.eu/articles/environmental-migrants-up-to-1-billion-by-2050/ (last accessed 1 June 2023).

16. Nick Watts et al., 'The 2018 report of the Lancet Countdown on health and climate change: Shaping the health of nations for centuries to come', *The Lancet* 392 (2018), pp. 2479–514, https://doi.org/10.1016/S0140-6736(18)32594-7 (last accessed 17 June 2023).

17. Lynne Jones, 'Letter to a woman on the Falklands Victory Parade, October 1982', in B. Harford and S. Hopkins (eds.), *Greenham Common: Women at the Wire* (London: Women's Press, 1984), pp. 74–7.

2. I WISH THERE WAS SOMEWHERE WE COULD MEET

1. Martin Luther King, 'Letter from Birmingham Jail (1963)', in Bob Blaisdell (ed.), *Essays on Civil Disobedience* (New York: Dover Publications, 2016), pp. 132–48.
2. 'Birmingham Campaign, April 3, 1963 to May 10, 1963', Stanford University, The Martin Luther King Jr Research and Education Institute, https://kinginstitute.stanford.edu/encyclopedia/birmingham-campaign (last accessed 1 June 2023).
3. 'Alabama clergymen's letter to Dr Martin Luther King Jr', 12 April 1963, https://web.archive.org/web/20181229055408/https://moodle.tiu.edu/pluginfile.php/57183/mod_resource/content/1/StatementAnd ResponseKingBirmingham1.pdf (last accessed 1 June 2023).
4. King, 'Letter from Birmingham Jail', pp. 133–5.
5. The terms global warming, climate change, and climate crisis are all used to describe the increased global temperature and consequent changes in climate brought about by increasing greenhouse gas emissions. My own preference is 'climate crisis' but I have used the term used by the author/ text/ report that I am quoting, or to which I am referring; hence the variation through this book.
6. *An Inconvenient Truth* was a documentary film made in 2006, directed by Davis Guggenheim, centred on a slide show by former United States Vice President Al Gore. The slide show was part of Gore's campaign to educate people on climate change and global warming. The film won an Academy Award for best documentary feature and has been widely viewed across the globe. A sequel was released in 2017.
7. Erica Chenoweth, *Civil Resistance: What Everyone Needs to Know* (New York: Oxford University Press, 2021), p. 114.
8. Extinction Rebellion, 'Act now', https://extinctionrebellion.uk/act-now/ (last accessed 1 June 2023).
9. Erica Chenoweth, 'Questions, Answers, and Some Cautionary Updates Regarding the 3.5% Rule', Carr Centre discussion paper, Spring 2020,

p. 1, https://carrcenter.hks.harvard.edu/publications/questions-answers-and-some-cautionary-updates-regarding-35-rule (last accessed 18 October 2023).

10. Ibid., p. 3.

11. Erica Chenoweth, 'People power', in Greta Thunberg (ed.), *The Climate Book* (London: Allen Lane, 2022), p. 367.

12. J. Liddington, *The Long Road to Greenham: Feminism and Anti-Militarism in Britain Since 1820* (London: Virago, 1989), pp. 240–1.

13. Ibid., p. 245.

14. Bea Campbell, *The Iron Ladies: Why Do Women Vote Tory?* (London: Virago, 1987), pp. 126, 130; cited in Liddington, *The Long Road to Greenham*, p. 253.

15. James and Ruby, 'Cultural roadblocks', in *This is Not a Drill: An Extinction Rebellion Handbook* (London: Penguin, 2019), pp. 115–19 (p. 115).

16. Tiana Jacout, Robin Boardman and Liam Geary Baulch, 'Building an action', in *This is Not a Drill*, pp. 109–11 (p. 109).

17. James O'Brien (@mrjamesob), Twitter, 17 October 2019, 8:01 AM, https://twitter.com/mrjamesob/status/1184726107104448512 (last accessed 1 June 2023).

18. Extinction Rebellion, 'About us' (2023), https://extinctionrebellion. uk/the-truth/about-us/ (last accessed 1 June 2023).

19. Jasper Jackson, 'Extinction Rebellion may enrage commuters, but it doesn't rely on majority support', *New Statesman*, 17 October 2019, https://www.newstatesman.com/politics/2019/10/extinction-rebel-lion-may-enrage-commuters-it-doesnt-rely-majority (last accessed 1 June 2023).

20. Michael Randle, *Civil Resistance* (London: Fontana, 1994), p. 189.

21. Author interview with Rupert Read, London, 22 March 2022.

22. Rupert Read and Roger Hallam, 'How political change works', Podcast, 21 January 2022, https://www.youtube.com/watch?v=-ehjc4r7b_k (last accessed 1 June 2023).

23. Rupert Read, 'What next on climate? The need for a new moderate flank', *Perspectiva*, 6 October 2021, https://systems-souls-society.com/what-next-on-climate-the-need-for-a-moderate-flank/ (last accessed 1 June 2023).

24. Author interview with Rupert Read, London, 22 March 2022.

25. Rupert Read, 'Why Insulate Britain needs a more positive strategy', Greenworld, 22 September 2021, https://greenworld.org.uk/article/why-insulate-britain-needs-more-positive-strategy (last accessed 1 June 2023).

26. Novara Media, 'Ink thrown at Insulate Britain protesters', 30 October 2021, https://novara.media/budget20212 (last accessed 2 June 2023).

27. Damien Gayle et al., 'Why do so many people hate Insulate Britain? Inside the controversial protest movement', Video, *The Guardian*, 17 November 2021, https://www.theguardian.com/global/video/2021/nov/17/why-do-so-many-people-hate-insulate-britain-inside-the-controversial-protest-movement (last accessed 2 June 2023).

28. Sam Petherick, 'Woman begs protesters to end blockade so she can follow her sick mum to hospital', *Metro*, 4 October 2021, https://metro.co.uk/2021/10/04/insulate-britain-woman-pleads-with-protesters-to-get-sick-mum-to-hospital-15359388/ (last accessed 2 June 2023).

29. Mike Galsworthy, 'Unbreak the planet with Mike Galsworthy: Episode 2', Podcast, The London Economic, 5 October 2021, https://www.youtube.com/watch?v=PWjtW-wnqXM (last accessed 2 June 2023).

30. Author interview with Rowan Tilly, Zoom, 28 February 2023.

31. Rowan Tilly, 'Ethics & nonviolence when blocking roads or motorway roundabout junctions/slip roads antinomy and disavowal: Notes & a point of view', unpublished paper.

32. 'Statement from Insulate Britain: We must acknowledge we have failed', Insulate Britain, 22 February 2022, https://insulatebritain.com/2022/02/07/statement-from-insulate-britain-we-must-acknowledge-we-have-failed/ (last accessed 2 June 2023).

33. Manish Pandey, 'Just Stop Oil: Why protesters are tying themselves to goalposts', BBC News, 21 March 2022, https://www.bbc.co.uk/news/newsbeat-60795041 (last accessed 2 June 2023).

34. Just Stop Oil press release, 'Breaking: Just Stop Oil coalition blocks 10 critical oil facilities to demand an end to new oil and gas', 1 April 2022, https://juststopoil.org/2022/04/01/breaking-just-stop-oil-coalition-blocks-7-critical-oil-facilities-to-demand-an-end-to-new-oil-and-gas/ (last accessed 2 June 2023).

35. Zack Sharf, '"Don't Look Up" comes to life: Climate activist battles "Good Morning Britain" anchor live on air', *Variety*, 14 April 2022, https://variety.com/2022/tv/news/dont-look-up-viral-climate-activist-battles-tv-news-anchor-1235232805/ (last accessed 2 June 2023).

36. Author interview with Chris Newman, London, 12 April 2022.

37. Mike Gill et al., 'We need health warning labels on points of sale of fossil fuels', *BMJ Opinion*, 31 March 2020, https://blogs.bmj.com/bmj/2020/03/31/we-need-health-warning-labels-on-points-of-sale-of-fossil-fuels/ (last accessed 2 June 2023).

38. Lynne Jones, 'Breaking barriers', *New Statesman*, October 1983, p. 14.

39. 26-year-old Stephen Waldorf was shot and seriously injured by armed police officers in London in January 1983 when he was mistaken for an escaped criminal. The ensuing public outcry resulted in new guidelines and training for the use of firearms by British police.

40. Paul Johnson was an English journalist, columnist and historian.

3. WHEN DO WE NEED TO BREAK DOWN THE DOOR?

1. Robert Reid, *The Peterloo Massacre* (Portsmouth, NH: Heinemann, 1989).

2. Ibid., p. 148.

3. Samuel Bamford, *Passages in the Life of a Radical, and Early Days*, vol. 2 (London: T. Fisher Unwin, 1841), p. 143.

4. E. P. Thompson, *The Making of the English Working Class* (London: Penguin, 1980), p. 747.

5. Shirin Hirsch, 'Protest and Peterloo: The story of 16 August 1819', People's History Museum, March 2019, https://phm.org.uk/protest-and-peterloo-the-story-of-16-august-1819/ (last accessed 2 June 2023).

6. Thompson, *Making of the English Working Class*, pp. 660–780.

7. 'The Tolpuddle Martyrs: The origins of trade unionism', *The Gazette*, https://www.thegazette.co.uk/all-notices/content/100553 (last accessed 2 June 2023).

8. Stewart Burns, 'Montgomery bus boycott', Encyclopedia of Alabama, 9 June 2008, updated 27 March 2023, https://encyclopediaofalabama.org/article/montgomery-bus-boycott/ (last accessed 2 June 2023).

9. John Rawls, 'The justification of civil disobedience', in Samuel Freeman

(ed.), *John Rawls: Collected Papers* (Cambridge, MA: Harvard University Press, 1999), pp. 176–89.

10. ICRC Database, 'Customary IHL: Practice relating to Rule 155, Defence of Superior Orders', https://ihl-databases.icrc.org/en/customary-ihl/v2/rule155?country=gb (last accessed 2 June 2023).

11. ICRC Database, Treaties, States Parties and Commentaries, 'Protocol Additional to the Geneva Conventions of 12 August 1949, and relating to the Protection of Victims of International Armed Conflicts (Protocol I), 8 June 1977, Article 48—Basic Rule', https://ihl-databases.icrc.org/en/ihl-treaties/api-1977/article-48 (last accessed 2 June 2023).

12. 'The Genocide Act 1969', The National Archives, https://www.legislation.gov.uk/ukpga/1969/12/2001-09-01 (last accessed 2 June 2023).

13. Molly Rush, 'A 40th anniversary Plowshares Eight reflection from Molly Rush', The Nuclear Resister, 9 September 2020, https://www.nukeresister.org/2020/09/09/a-40th-anniversary-plowshares-eight-reflection-from-molly-rush/ (last accessed 2 June 2023).

14. See 'Silo Pruning Hooks', Archive today, https://archive.ph/0tOBk (last accessed 4 September 2023); and 'A history of direct disarmament actions', http://coat.ncf.ca/our_magazine/links/issue42/articles/a_history_of_direct_disarmament.htm (last accessed 2 June 2023).

15. Hugh O'Shaughnessy and Matthew Brace, 'Campaigners face jail for raid on military jet', *The Independent*, 20 July 1996, https://www.independent.co.uk/news/uk/home-news/campaigners-face-jail-for-raid-on-military-jet-1329683.html (last accessed 2 June 2023). Joanna Wilson later changed her name to Jo Blackman.

16. Andrea Needham, *The Hammer Blow: How Ten Women Disarmed a War Plane* (London: Peace News Press, 2016).

17. Andreas Marcou, 'Violence, communication, and civil disobedience', *Jurisprudence* 12, no. 4 (2021), pp. 491–511, https://doi.org/10.1080/20403313.2021.1921494 (last accessed 18 August 2023).

18. Jessica Reznicek and Ruby Montoya, 'Why we acted', *Via Pacis: The Voice of the Des Moines Catholic Worker Community* 41, no. 3 (October 2017), pp. 1, 3.

19. Philip Joens, 'Iowa climate activist sentenced to eight years in federal prison for Dakota Access Pipeline sabotage', *Des Moines Register*,

30 June 2021, https://eu.desmoinesregister.com/story/news/crime-and-courts/2021/06/30/iowa-activist-jessica-reznicek-sentenced-dakota-access-pipeline-sabotage-catholic-workers/7808907002/ (last accessed 2 June 2023).

20. Julia Shipley, 'The long legal saga of DAPL arsonist Ruby Montoya is coming to an end', Grist, 21 September 2022, https://grist.org/protest/ruby-montoya-dakota-access-pipeline/ (last accessed 2 June 2023).

21. Fern Riddell, 'Sanitising the suffragettes', *History Today* 68, no. 2 (February 2018), https://www.historytoday.com/history-matters/sanitising-suffragettes (last accessed 2 June 2023).

22. 'July 19th, 1912: From the archives', *Irish Times*, 19 July 2010, https://www.irishtimes.com/opinion/july-19th-1912-from-the-archives-1.623855 (last accessed 2 June 2023).

23. Book interview with Fern Riddell: 'Can we call the suffragettes terrorists? "Absolutely," says Fern Riddell', History Extra, 7 December 2020, https://www.historyextra.com/period/edwardian/books-interview-with-fern-riddell-can-we-call-the-suffragettes-terrorists-absolutely/ (last accessed 2 June 2023).

24. Simon Webb, *The Suffragette Bombers: Britain's Forgotten Terrorists* (Barnsley: Pen and Sword, 2014).

25. C. J. Bearman, 'An examination of suffragette violence', *English Historical Review* 120, no. 486 (April 2005), pp. 365–97, https://academic.oup.com/ehr/article-abstract/120/486/365/396169 (last accessed 2 June 2023).

26. Fern Riddell, *Death in Ten Minutes: The Forgotten Life of Radical Suffragette Kitty Marion* (London: Hodder and Stoughton, 2018), pp. 117–21.

27. Fern Riddell, 'Suffragettes, violence and militancy', British Library, 6 February 2018, https://www.bl.uk/votes-for-women/articles/suffragettes-violence-and-militancy#footnote13 (last accessed 2 June 2023).

28. Emmeline Pankhurst, '"Freedom or death": Speech delivered in Hartford, Connecticut on November 13, 1913', *The Guardian*, 27 April 2007, https://www.theguardian.com/theguardian/2007/apr/27/greatspeeches1 (last accessed 2 June 2023).

29. WSPU Seventh Annual Report (1913), p. 16, quoted in Bearman, 'Examination of suffragette violence', p. 375.

30. *The Times*, 12 June 1914, quoted in Bearman, 'Examination of suffragette violence', p. 394.

31. Philip Snowden, 'The present position of woman suffrage', *The Englishwoman* 60 (1913), pp. 244–5. The 1914 speech to the Christian Fellowship of 13 June, quoted in *The Birmingham Daily Mail* on 15 June, is quoted in Bearman, 'Examination of suffragette violence', p. 395.

32. Extinction Rebellion, 'Can property damage be nonviolent?', Video, March 2021, https://www.youtube.com/watch?v=FU4YY9veisU (last accessed 2 June 2023).

33. Zoe Blackler, 'Magistrates rule Extinction Rebellion protestor has "lawful excuse" for criminal damage due to climate emergency', Extinction Rebellion, 1 November 2019, https://extinctionrebellion.uk/2019/11/01/Magistrates-rule-extinction-rebellion-protestor-has-lawful-excuse-for-criminal-damage-due-to-climate-emergency/ (last accessed 2 June 2023).

34. Sandra Laville and agencies, 'Extinction Rebellion founder cleared over King's College protest', *The Guardian*, 9 May 2019, https://www.theguardian.com/environment/2019/may/09/extinction-rebellion-founder-cleared-over-kings-college-protest (last accessed 2 June 2023).

35. David Lambert, 'A defence of criminal damage', RSA, 13 May 2021, https://www.thersa.org/comment/2021/05/a-defence-of-criminal-damage (last accessed 2 June 2023).

36. Press release: 'The XR activists who took on oil giant Shell—and won', Extinction Rebellion, 23 April 2021, https://extinctionrebellion.uk/2021/04/23/breaking-the-Extinction Rebellion-activists-who-took-on-oil-giant-shell-and-won/ (last accessed 2 June 2023).

37. 'Extinction Rebellion: Jury acquits protesters despite judge's direction', BBC News, 23 April 2021, https://www.bbc.co.uk/news/uk-england-london-56853979 (last accessed 2 June 2023).

38. Tess de la Mare and Dawn Limbu, 'Extinction Rebellion co-founder fined for smashing Barclays window', BBC News, 6 January 2023, https://www.bbc.co.uk/news/uk-england-gloucestershire-64193016 (last accessed 2 June 2023).

39. Press release: '"Better broken windows than broken promises"— Extinction Rebellion women break windows at Barclays HQ in Canary

Wharf', Extinction Rebellion, 7 April 2021, https://extinctionrebellion.uk/2021/04/07/breaking-better-broken-windows-than-broken-promises-extinction-rebellion-women-break-windows-at-barclays-hq-in-canary-wharf/ (last accessed 2 June 2023).

40. Chris Skinner, 'A banker responds to Extinction Rebellion's call', Chris Skinner's Blog, 10 February 2022, https://thefinanser.com/2022/02/a-banker-responds-to-extinction-rebillions-call (last accessed 2 June 2023).

41. Gail Bradbrook, 'How can breaking bank windows be justified?', Chris Skinner's Blog, 4 April 2022, https://thefinanser.com/2022/04/breaking-bank-windows-is-justified.html/ (last accessed 2 June 2023).

42. Damien Gayle, 'Just Stop Oil activists throw soup at Van Gogh's Sunflowers', *The Guardian*, 14 October 2022, https://www.theguardian.com/environment/2022/oct/14/just-stop-oil-activists-throw-soup-at-van-goghs-sunflowers (last accessed 2 June 2023).

43. See Michael Mezz (@michaelmezz), Twitter, 18 October 2022, 2:39 AM, https://twitter.com/michaelmezz/status/1582184473252098049 (accessed 18 August 2023).

44. Peggy Jones, 'Why I changed my mind about the attack on Van Gogh's Sunflowers', Medium, 21 October 2022, https://medium.com/politically-speaking/why-i-changed-my-mind-about-the-attack-on-van-goghs-sunflowers-b3c0e718f39a (last accessed 2 June 2023).

45. 'Two Buffalo police officers shove a man to the ground in front of City Hall', WFBO, 4 June 2020, https://www.youtube.com/watch?v=QSBZGv5wzK4 (last accessed 9 July 2023).

46. Tristan Cork, 'Row breaks out as Merchant Venturer accused of "sanitising" Edward Colston's involvement in slave trade', Bristol Live, 23 August 2018, https://web.archive.org/web/20200607220922/https://www.bristolpost.co.uk/news/bristol-news/row-breaks-out-merchant-venturer-1925896 (last accessed 9 July 2023).

47. David Olusoga, 'The toppling of Edward Colston's statue is not an attack on history. It is history', *The Guardian*, 8 June 2020, https://www.theguardian.com/commentisfree/2020/jun/08/edward-colston-statue-history-slave-trader-bristol-protest (last accessed 15 October 2023).

48. Ellie Pipe, 'Report examines police response when Colston's statue was toppled', B24/7, 11 March 2021, https://www.bristol247.com/news-and-features/news/report-examines-police-response-when-colstons-statue-was-toppled/ (last accessed 2 June 2023).

49. Conor Gogarty, 'Priti Patel says toppling of Colston statue is "utterly disgraceful"—but Piers Morgan hits back', Bristol Live, 7 June 2020, https://www.bristolpost.co.uk/news/bristol-news/priti-patel-says-toppling-colston-4202300 (last accessed 2 June 2023).

50. W. E. Gladstone, *The Slavonic Provinces of the Ottoman Empire* (London, 1877), cited in Roland Quinault, 'Gladstone and slavery', *Historical Journal* 52, no. 2 (June 2009), pp. 363–83.

51. Charles Dickens, 'To Miss Burdett-Coutts, 4 October 1857', in Graham Storey and Kathleen Mary Tillotson (eds.), *The British Academy/The Pilgrim Edition of the Letters of Charles Dickens, Vol. 8: 1856–1858* (Oxford: Oxford University Press, 1995), https://dx.doi.org/10.1093/oseo/instance.00161597 (last accessed 2 June 2023).

52. Angelique Retief, 'Housing in 2020: Through the lens of racial justice', Black South West Network, 19 June 2020, https://www.blacksouthwestnetwork.org/blog/housing-in-2020-through-the-lens-of-racial-justice/19/6/2020 (accessed 2 June 2023).

53. 'Pressure grows over Bomber Harris statue', Peace Pledge Union, 16 June 2020, https://www.ppu.org.uk/news/pressure-grows-over-bomber-harris-statue (last accessed 2 June 2023).

54. Peter Walker, '"We cannot edit our past": Boris Johnson's statue tweets explained', *The Guardian*, 12 June 2020, https://www.theguardian.com/politics/2020/jun/12/we-cannot-edit-our-past-boris-johnsons-statue-tweets-explained (last accessed 2 June 2023).

55. 'Colston statue should be displayed in museum, public says', BBC News, 3 February 2022, https://www.bbc.co.uk/news/uk-england-bristol-60247092 (last accessed 2 June 2023).

56. 'Bristol slave trader Edward Colston statue toppling: Four on trial', BBC News, 13 December 2021, https://www.bbc.co.uk/news/uk-england-bristol-59611875 (last accessed 2 June 2023).

57. 'Edward Colston: Protester called statue toppling an "act of love"', BBC News, 16 December 2021, https://www.bbc.co.uk/news/uk-england-bristol-59686345 (last accessed 2 June 2023).

58. 'Edward Colston statue: Four cleared of criminal damage', BBC News, 5 January 2022, https://www.bbc.co.uk/news/uk-england-bristol-59727161 (last accessed 2 June 2023).

59. Rhian Graham, 'I'm one of the Colston Four. Our victory confirms the power and value of protest', *The Guardian*, 6 January 2022, https://www.theguardian.com/commentisfree/2022/jan/06/colston-four-victory-racial-justice-history (last accessed 2 June 2023).

60. Lynne Jones, 'In the eye of the storm', *New Statesman*, 16 December 1983, pp. 8–9. This piece was written one month after the cruise missiles had actually been deployed at Greenham Common airbase.

61. On 14 November 1983, Defence Secretary Michael Heseltine announced in parliament the arrival of the first cruise missiles at Greenham Common. James Cameron was a journalist who sometimes wrote about Greenham women in *The Guardian*.

4. WHO ARE YOUR LEADERS?

1. Doctors for Extinction Rebellion, Letter to Rishi Sunak, Chancellor of the Exchequer, 7 April 2022.

2. Lozza's Lock-In, 'Two-tiered policing and doomsday eco-loons', Reclaim Party, Facebook, 8 April 2022, https://www.facebook.com/thereclaimpartyofficial/videos/398002388468967/ (last accessed 3 March 2023).

3. Henry Goodwin, 'Insulate Britain success: Brits say insulation is best way to curb Russian gas use', The London Economic, 6 April 2022, https://www.thelondoneconomic.com/politics/insulate-britain-success-brits-say-insulation-is-best-way-to-curb-russian-gas-use-318544/ (last accessed 3 March 2023).

4. Press release: 'Secretary-General warns of climate emergency, calling Intergovernmental Panel's Report "a file of shame", while saying leaders "are lying", fuelling flames', United Nations Meetings Coverage and Press Releases, 4 April 2022, https://press.un.org/en/2022/sgsm21228.doc.htm (last accessed 3 June 2023).

5. Ibid.

6. Author interview with Alice Clack, Zoom, 12 December 2022.

7. Health for Extinction Rebellion website (formerly Doctors for Extinction Rebellion), https:// healthforxr.com/ (last accessed 22 July 2023).

8. Royal College of Psychiatrists, 'RCPsych declares a climate and ecological emergency', RCPsych, 6 May 2021, https://www.rcpsych.ac. uk/news-and-features/latest-news/detail/2021/05/05/rcpsych-declares-a-climate-and-ecological-emergency (last accessed 3 March 2023).

9. Medact, https://www.medact.org/ (last accessed 3 March 2023).

10. 'XR in your area', Extinction Rebellion, https://rebellion.global/groups/#countries (last accessed 9 September 2023).

11. Email to author from Helen Angel, 26 July 2023.

12. 'UK rebel hive', Extinction Rebellion, https://organism.extinctionrebellion.uk/?id=35 (last accessed 3 March 2023).

13. Jo Freeman (aka Joreen), 'The tyranny of structurelessness' [1970], Jo Freeman.com, https://www.jofreeman.com/joreen/tyranny.htm (last accessed 3 June 2023).

14. Richard Norton-Taylor, 'MI5 put union leaders and protesters under surveillance during Cold War', *The Guardian*, 6 October 2009, https://www.theguardian.com/uk/2009/oct/06/mi5-union-leaders-surveillance (last accessed 3 June 2023).

15. Chenoweth, *Civil Resistance*, pp. 121–2.

16. Marshall B. Rosenberg, *Nonviolent Communication: A Language of Compassion* (Del Mar, CA: Puddle Dancer Press, 1999).

17. Author interview with Mark Wingfield, Zoom, 18 December 2022.

18. 2019 general election results, https://electionresults.parliament.uk/election/2019–12–12/Results/Location/Constituency/St%20Ives/ (accessed 14 September 2023).

19. 'How the 2019 election results could have looked with proportional representation', Electoral Reform Society, 13 December 2019, https://www.electoral-reform.org.uk/how-the-2019-election-results-could-have-looked-with-proportional-representation/ (last accessed 3 March 2023).

20. '2015 general election results, St Ives', UK Parliament, https://electionresults.parliament.uk/election/2015–05–07/Results/Location/Constituency/St%20Ives/ (last accessed 3 March 2023).

21. Briefing paper 'UK Election Statistics: 1918–2023: A century of elections', House of Commons Library, 9 August 2023, https://commonslibrary.parliament.uk/research-briefings/cbp-7529/ (last accessed 18 October 2023).

22. George Washington, *George Washington's Farewell Address* (Carlisle: Applewood Books, 1999).

23. Frans de Waal, *The Bonobo and the Atheist: In Search of Humanism among the Primates* (New York: W. W. Norton, 2013), pp. 228–40.

24. Oxfam International, 'Inequality kills', Oxfam Briefing Paper, January 2022, https://www.oxfam.org/en/research/inequality-kills (last accessed 4 June 2023).

25. Oxfam International, 'Survival of the richest', Oxfam Briefing Paper, January 2023, https://www.oxfam.org/en/research/survival-richest (last accessed 4 June 2023).

26. Adam Bychawski, 'Fears over cost of living "solutions" proposed by Truss-backed think tanks', Open Democracy, 1 September 2022, https://www.opendemocracy.net/en/liz-truss-think-tanks-cost-of-living-iea/ (last accessed 4 April 2023).

27. 'Boris Johnson makes more than £1m from speeches since leaving office', BBC News, 14 December 2022, https://www.bbc.co.uk/news/uk-politics-63975286 (last accessed 4 June 2023).

28. Joseph Rowntree Foundation, 'UK poverty 2023: The essential guide to understanding poverty in the UK', January 2023, https://www.jrf.org.uk/report/uk-poverty-2023#key-findings (last accessed 4 June 2023).

29. Kevin MacKay, 'The ecological crisis is a political crisis', MAHB Blog, 25 September 2018, https://mahb.stanford.edu/blog/ecological-crisis-political-crisis/ (last accessed 4 April 2023).

30. 'World has "gambled on fossil fuels and lost", warns Guterres', UN News, 17 June 2022, https://news.un.org/en/story/2022/06/1120662 (last accessed 4 June 2023).

31. David Runciman, *How Democracy Ends* (London: Profile Books, 2018), p. 132.

32. Benjamin Rush, 'Address to the people of the United States, American Museum, January 1787', in John P. Kaminski et al. (eds.), *The Documentary History of the Ratification of the Constitution Digital Edition*, (Charlottesville, VA: University of Virginia Press, 2009). Original source: *Commentaries on the Constitution, Vol. XIII: Commentaries on the Constitution*, No. 1, https://archive.csac.history.wisc.edu/Benjamin_Rush.pdf (last accessed 4 June 2023).

33. Christopher Blackwell, 'An introduction to classical Athenian democracy—Overview', Discussion series, Athenian Law Lectures, Center for Hellenic Studies, Harvard University, 2 November 2020, https://chs.harvard.edu/discussion-series-athenian-law-lectures-8/ (last accessed 5 June 2023).

34. Rutger Bregman, *Humankind: A Hopeful History* (London: Bloomsbury, 2020), pp. 300–8.

35. Michela Palese, 'The Irish abortion referendum: How a Citizens' Assembly helped to break years of political deadlock', Electoral Reform Society, 29 May 2018, https://www.electoral-reform.org.uk/the-irish-abortion-referendum-how-a-citizens-assembly-helped-to-break-years-of-political-deadlock/ (last accessed 5 June 2023).

36. Ronan McGreevy, 'Why did Citizens' Assembly take liberal view on abortion?', *The Irish Times*, 30 June 2017, https://www.irishtimes.com/news/social-affairs/why-did-citizens-assembly-take-liberal-view-on-abortion-1.3138280 (last accessed 5 June 2023).

37. Dominik Hierlemann, Anna Renkamp and Robert Vehrkamp, 'The Ostbelgien model: Institutionalising deliberative democracy', BertelsmanStiftung, 7 March 2022, https://www.bertelsmann-stiftung.de/en/our-projects/democracy-and-participation-in-europe/shortcut-archive/shortcut-7-the-ostbelgien-model (last accessed 5 June 2023).

38. Climate Assembly UK, 'The path to net zero', 10 September 2020, https://www.climateassembly.uk/news/uk-path-net-zero-must-be-underpinned-education-choice-fairness-and-political-consensus-urges-climate-assembly/ (last accessed 5 June 2023).

39. Climate Assembly UK, 'The path to net zero: Executive summary', 2020, pp. 4, 5, https://www.climateassembly.uk/report/read/final-report-exec-summary.pdf (last accessed 5 June 2023).

40. Ibid., p. 12.

41. Ibid., pp. 28–9.

42. OECD, 'Innovative citizen participation and new democratic institutions: Catching the deliberative wave', OECD Publishing, Paris, June 2020, https://www.oecd.org/gov/innovative-citizen-participation-and-new-democratic-institutions-339306da-en.htm (last accessed 5 April 2023).

43. Author interview with Dr Jan Power, Penzance, 30 January 2023.

44. Author interview with Myghal Ryual, Falmouth, 2 May 2023.

45. 'Council Tax premium on second homes approved by Cabinet', Cornwall Council, 16 December 2022, updated 29 June 2023, https://www.cornwall.gov.uk/council-news/council-budgets-and-economy/council-tax-premium-on-second-homes-approved-by-cabinet/ (last accessed 5 June 2023).

46. Lynne Jones, 'Shut up and listen', *New Statesman*, 2 March 1984, pp. 10–11.

47. This is equivalent to just over £100 today.

5. LAW OR JUSTICE?

1. ''We are united': More than 200 health journals call for emergency action on climate change', Sky News, 6th September 2021, https://news.sky.com/story/we-are-united-more-than-200-health-journals-call-for-emergency-action-on-climate-change-12400336 (last accessed 19 November 2023).

2. Lancet Planetary Health Editorial, 'A role for provocative protest', *The Lancet Planetary Health* 6, no. 11 (November 2022), e846, https://www.thelancet.com/journals/lanplh/article/PIIS2542-5196(22)00287-X/fulltext (last accessed 22 July 2023).

3. 'Court finds XR scientists not guilty for action at BEIS—while charges are dropped against Shell HQ occupation defendants', Extinction Rebellion, 21 October 2022, https://extinctionrebellion.uk/2022/10/21/breaking-court-finds-xr-scientists-not-guilty-for-action-at-beis-while-charges-are-dropped-against-shell-hq-occupation-defendants/ (last accessed 5 June 2023).

4. Kevin Rawlinson, 'Jury clears Extinction Rebellion activists who targeted commuters', *The Guardian*, 10 December 2021, https://www.theguardian.com/uk-news/2021/dec/10/jury-clears-extinction-rebellion-activists-who-targeted-commuters (last accessed 5 June 2023).

5. PA News Agency, 'Acquittal of activists shows public taking climate crisis "far more seriously"', *Bracknell News*, 14 January 2022, https://www.bracknellnews.co.uk/news/national/19849420.acquittal-activists-shows-public-taking-climate-crisis-far-seriously/ (last accessed 5 June 2023).

6. Press Summary, 'Director of Public Prosecutions (Respondent) v Ziegler and others (Appellants) [2021] UKSC 23 On appeal from: [2019] EWHC 71 (Admin), Supreme Court', 25 June 2021, https:// www.supremecourt.uk/press-summary/uksc-2019-0106.html (last accessed 5 June 2023).

7. Milan Rai, 'Ziegler: The full story behind the ground-breaking Supreme Court decision', *Peace News*, 20 June 2022, https://peace-news.info/blog/2022/ziegler-full-story-behind-ground-breaking-supreme-court-decision (last accessed 5 June 2023).

8. Press Summary: Director of Public Prosecutions (Respondent) v Ziegler and others (Appellants) [2021] UKSC 23 On appeal from: [2019] EWHC 71 (Admin), 'Background to the appeal', https://www.supremecourt.uk/press-summary/uksc-2019-0106.html (accessed 24 August 2023).

9. Milan Rai, 'Ziegler: Celebrating the Supreme Court decision one year on', *Peace News*, 1 June 2022, https://peacenews.info/node/10270/ziegler-celebrating-supreme-court-decision-one-year (last accessed 5 June 2023).

10. Zoe Blackler, 'First XR defendant acquitted under Supreme Court's "Ziegler" ruling', Extinction Rebellion, 21 July 2021, https://extinctionrebellion.uk/2021/07/21/first-xr-defendant-acquitted-under-supreme-courts-ziegler-ruling/ (last accessed 5 June 2023).

11. Calum Paton, 'The Colston Four: What does the verdict mean for the UK justice system?', The Speaker, 6 January 2022, https://speakerpolitics.co.uk/inspiringeducation/explainer-topics/analysis-parent/analysis/the-colston-four-what-does-the-verdict-mean-for-the-uk-justice-system/ (last accessed 5 June 2023).

12. Sam Tobin, 'Police acted unlawfully over Everard vigil, court rules', *Law Society Gazette*, 11 March 2022, https://www.lawgazette.co.uk/news/police-acted-unlawfully-over-everard-vigil-court-rules/5111836.article (last accessed 5 June 2023).

13. Sam Tobin, 'News focus: Colston Four—AG accused of subverting jury system', *Law Society Gazette*, 4 July 2022, https://www.lawgazette.co.uk/news-focus/news-focus-colston-four-ag-accused-of-subverting-jury-system/5113003.article (last accessed 5 June 2023).

14. Press Summary: 'Case no: 202201151 B3: Attorney General's reference on a point of law, No. 1 of 2022 (pursuant to section 36 of the Criminal Justice Act 1972)', Courts and Tribunals Judiciary, 28 September 2022, https://www.judiciary.uk/wp-content/uploads/2022/09/AG-Ref-Colston-Four-summary-280922.pdf (last accessed 5 June 2023).

15. 'Court finds XR scientists not guilty for action at BEIS—while charges are dropped against Shell HQ occupation defendants', Extinction Rebellion, 21 October 2022, https://extinctionrebellion.uk/2022/10/21/breaking-court-finds-xr-scientists-not-guilty-for-action-at-beis-while-charges-are-dropped-against-shell-hq-occupation-defendants/ (last accessed 5 June 2023).

16. 'Statement: Judge rules the moral thing to do in these times is to sit on the M25', Insulate Britain, 11 October 2022, https://insulatebritain.com/2022/10/11/statement-Judge-rules-the-moral-thing-to-do-in-these-times-is-to-sit-on-the-m25/ (last accessed 5 June 2023).

17. 'Breaking: Four Insulate Britain supporters vindicated after jury returns unanimous not guilty verdict', Insulate Britain, 11 January 2023, http://insulatebritain.com/2023/01/11/breaking-four-insulate-britain-supporters-vindicated-after-jury-returns-unanimous-not-guilty-verdict/ (last accessed 5 June 2023).

18. 'Breaking: Three Insulate Britain supporters vindicated after jury returns unanimous not guilty verdict', Insulate Britain, 9 December 2022, http://insulatebritain.com/2022/12/09/breaking-three-insulate-britain-supporters-vindicated-after-jury-returns-unanimous-not-guilty-verdict/ (last accessed 5 June 2023).

19. 'Four scientists for XR vindicated by Crown Court following appeal', Extinction Rebellion, 10 February 2023, https://extinctionrebellion.uk/2023/02/10/four-scientists-for-xr-vindicated-by-crown-court-following-appeal/ (last accessed 5 June 2023).

20. 'Statement on Canning Town sentencing', Extinction Rebellion, 18 March 2022, https://extinctionrebellion.uk/2022/03/18/statement-on-canning-town-sentencing/ (last accessed 5 June 2023).

21. 'Three Just Stop Oil supporters released after 109 days in prison without trial', Just Stop Oil, 22 February 2023, https://juststopoil.org/2023/02/22/three-just-stop-oil-supporters-released-after-

109-days-in-prison-without-trial/ and https://juststopoil.org/news-press/ (last accessed 5 June 2023).

22. Zoe Blackler, 'Judge waives punishment for "noble" protester guilty of blocking traffic', Extinction Rebellion', 20 May 2021, https://extinctionrebellion.uk/2021/05/20/judge-waives-punishment-for-noble-protester-guilty-of-blocking-traffic/ (last accessed 5 June 2023).

23. Nina Lloyd, 'Judge praises Insulate Britain protesters despite fining them', Yahoo!Finance, 12 April 2022, https://ca.finance.yahoo.com/news/judge-praises-insulate-britain-protesters-183341590.html (last accessed 5 June 2023).

24. 'Breaking: Four Insulate Britain supporters vindicated after jury returns unanimous not guilty verdict', Insulate Britain, 11 January 2023, http://insulatebritain.com/2023/01/11/breaking-four-insulate-britain-supporters-vindicated-after-jury-returns-unanimous-not-guilty-verdict/ (last accessed 5 June 2023).

25. C. Douzinas, S. Homewood and R. Warrington, 'The shrinking scope for public protest', *Index on Censorship* 17, no. 8 (1988), pp. 12–15, https://doi.org/10.1080/03064228808534499 (last accessed 11 July 2023).

26. Tim Wilson and Richard Walton, 'Extremism rebellion: A review of ideology and tactics', Policy Exchange, July 2019, https://policyexchange.org.uk/wp-content/uploads/2019/07/Extremism-Rebellion.pdf (last accessed 2 September 2023).

27. Adam Bychawski, 'Revealed: Policing bill was dreamed up by secretive oil-funded think tank', Open Democracy, 15 June 2022, https://www.opendemocracy.net/en/dark-money-investigations/policing-bill-policy-exchange-exxonmobil-lobbying/ (last accessed 5 June 2023).

28. Vikram Dodd and Jamie Grierson, 'Terrorism police list Extinction Rebellion as extremist ideology', *The Guardian*, 10 January 2020, https://www.theguardian.com/uk-news/2020/jan/10/xr-extinction-rebellion-listed-extremist-ideology-police-prevent-scheme-guidance (last accessed 5 June 2023).

29. 'Extinction Rebellion blocks News Corps Printworks and demands they "Free the Truth"', Extinction Rebellion, 4 September 2020, https://extinctionrebellion.uk/2020/09/04/breaking-extinction-rebel-

lion-blocks-news-corps-printworks-and-demands-they-free-the-truth/ (last accessed 5 June 2023).

30. 'Home Secretary speech at the Police Superintendents' Association conference', GOV.UK, 8 September 2020, https://www.gov.uk/government/speeches/home-secretary-speech-at-the-police-superintendents-association-conference (last accessed 5 June 2023).

31. C. Hickman et al., 'Climate anxiety in children and young people and their beliefs about government responses to climate change: A global survey', *Lancet Planetary Health* 5, no. 12 (2021), e863–73.

32. A. Sanson and M. Bellemo, 'Children and youth in the climate crisis', *BJPsych Bulletin* 45, no. 4 (August 2021), pp. 205–9, http://doi:10.1192/bjb.2021.16 (last accessed 5 June 2023).

33. Damien Gayle, 'Psychiatrists warn of police and crime bill's impact on young people', *The Guardian*, 15 January 2022, https://www.theguardian.com/uk-news/2022/jan/15/psychiatrists-warn-of-police-and-bills-impact-on-young-people (last accessed 5 June 2023).

34. Police Bill Alliance (@PoliceBillAll), Twitter, 17 January 2022, 4:02 PM, https://twitter.com/PoliceBillAll/status/1483107445979426819?s=20 (last accessed 5 June 2023).

35. Debate: 'Amendment 133A: Police, Crime, Sentencing and Courts Bill—*Report (6th Day)*—in the House of Lords at 9:15 pm on 17th January 2022', https://www.theyworkforyou.com/lords/?id=2022–01–17a.1419.5 (last accessed 5 June 2023).

36. Police Bill Alliance (@PoliceBillAll), Twitter, 17 January 2022, 7:40 PM, https://twitter.com/PoliceBillAll/status/1483162290182627336 (last accessed 2 September 2023).

37. James Tapsfield, 'Eco "hooligans" will be BANNED from chaining themselves to buildings and blocking roads under tough new measures in Queen's Speech', *Mail Online*, 10 May 2022, https://www.dailymail.co.uk/news/article-10801549/Queens-Speech-says-police-new-powers-tackle-eco-hooligans.html (last accessed 5 June 2023).

38. 'Global trends in climate change litigation: 2022 snapshot', LSE and Grantham Research Institute on Climate Change and the Environment, 30 June 2022, https://www.lse.ac.uk/granthaminstitute/publication/global-trends-in-climate-change-litigation-2022/ (last accessed 5 June 2023).

39. Greenham Women Against Cruise Missiles v. Reagan, 591 F. Supp. 1332 (S.D.N.Y. 1984), 31 July 1984, Justia US Law, https://law.justia.com/cases/federal/district-courts/FSupp/591/1332/2388846/ (last accessed 5 June 2023).

40. 'An trial like no other', Global Legal Action Network, https://youth-4climatejustice.org (last accessed 11 September 2023).

41. Author interview with Ben Cooper, Zoom, 23 January 2023.

42. Kate Connolly, '"Historic" German ruling says climate goals not tough enough', *The Guardian*, 29 April 2021, https://www.theguardian.com/world/2021/apr/29/historic-german-ruling-says-climate-goals-not-tough-enough (last accessed 5 June 2023).

43. Noni Shannon and Elisa de Witt, 'Urgenda Foundation v. Netherlands: Historic climate change decision upheld', Norton Rose Fulbright, December 2019, https://www.nortonrosefulbright.com/en-au/knowledge/publications/45dc4f83/urgenda-foundation-v-netherlands-historic-climate-change-decision-upheld (last accessed 5 June 2023).

44. Antoni Juhasz, 'The quest to defuse Guyana's carbon bomb', Wired, 20 December 2022, https://www.wired.com/story/the-quest-to-defuse-carbon-bomb-guyana/ (last accessed 5 June 2023).

45. Ibid.

46. G. Supran et al., 'Assessing ExxonMobil's global warming projections', *Science* 379, no. 6628 (13 January 2023), https://www.science.org/doi/10.1126/science.abk0063 (last accessed 5 June 2023).

47. Author interview with Melinda Janki, London, 27 February 2023.

48. Aliyah Elfar, 'Landmark climate change lawsuit moves forward as German judges arrive in Peru', Columbia Climate School: State of the Planet, 4 August 2022, https://news.climate.columbia.edu/2022/08/04/landmark-climate-change-lawsuit-moves-forward-as-german-Judges-arrive-in-peru/ (last accessed 5 June 2023).

49. Milieudefensie et al. v. Royal Dutch Shell plc., Climate Change Litigation Databases, http://climatecasechart.com/non-us-case/milieudefensie-et-al-v-royal-dutch-shell-plc/ (last accessed 5 June 2023).

50. 'Historic High Court ruling finds UK government's climate strategy "unlawful"', Client Earth, 18 July 2022, https://www.clientearth.org/latest/press-office/press/historic-high-court-ruling-finds-uk-government-s-climate-strategy-unlawful/ (last accessed 5 June 2023).

51. 'The Energy Charter Treaty', Global Justice Now, 1 February 2023, https://www.globaljustice.org.uk/resource/the-energy-charter-treaty (last accessed 5 June 2023).

52. The Small Island Developing States (SIDS) are a distinct group of thirty-nine States and 18 Associate Members of United Nations regional commissions that face unique social, economic and environmental vulnerabilities. They are located in the Caribbean, the Pacific, the Atlantic, the Indian Ocean and the South China Sea, https://www.un.org/ohrlls/content/about-small-island-developing-states (last accessed 12 July 2023).

53. Plan B: About, https://planb.earth/about/ (last accessed 25 August 2023).

54. Tim Crosland, 'Personal statement on the Supreme Court's Heathrow verdict', Plan B, 16 December 2020, https://planb.earth/supreme-court-betrays-young-and-global-south/ (last accessed 5 June 2023).

55. Author interview with Tim Crosland, London, 27 January 2023.

56. Open letter: 'Lawyers for 1.5°C—Humanity's lifeline', https://planb.earth/wp-content/uploads/2022/09/Lawyers-letter-FINAL-PDF.pdf (last accessed 2 September 2023).

57. Agencies, 'Climate activists guilty of smashing Barclays HQ windows spared jail', *The Guardian*, 27 January 2023, https://www.theguardian.com/world/2023/jan/27/climate-activists-guilty-smashing-barclays-hq-windows-escape-jail (last accessed 5 June 2023).

58. Lynne Jones, 'Changing ideas of authority', *New Statesman*, 20 November 1984, pp. 8–9.

6. HOW TO PREVENT HARM

1. 'Cruise missiles locked inside Greenham Base', *Jane's Defence Weekly*, 14 January 1984, p. 6.

2. Joan Smith, 'Officialdom grinds the peace camps', *New Statesman*, 14 February 1986.

3. Martin Levy, *Ban the Bomb! Michael Randle and Direct Action against Nuclear War* (Stuttgart: Ibidem Press, 2021), p. 9.

4. Christopher Driver, *The Disarmers: A Study in Protest* (London: Hodder and Stoughton, 1964).

5. Laurence Freedman, 'Thatcherism and defence', in Dennis Kavanagh and Anthony Seldon (eds.), *The Thatcher Effect* (Oxford: Clarendon Press, 1989), pp. 143–53.

6. Jones, 'Shut up and listen', pp. 10–11.

7. Liddington, *The Long Road to Greenham*, p. 280.

8. Gene Sharp, *The Politics of Nonviolent Action, Part Two: The Methods of Nonviolent Action* (Boston, MA: Porter Sargent, 1973), pp. 357–435.

9. Ibid., pp. 375–7.

10. Ibid., p. 377.

11. Gene Sharp, *The Politics of Nonviolent Action, Part Three: The Dynamics of Nonviolent Action* (Boston, MA: Porter Sargent, 1973), pp. 657–97.

12. Sharp, *Politics of Nonviolent Action, Part Two: The Methods*, p. 377.

13. Samuel J. May, 'Discourse delivered in the Church of the Messiah, Syracuse, 3 November 1850', *Anti-Slavery Bugle*, 5 April 1851. Quoted in Jane H. Pease and William H. Pease, 'Confrontation and abolition in the 1850s', *Journal of American History* 58, no. 4 (1972), pp. 923–37.

14. Sharp, *Politics of Nonviolent Action, Part Two: The Methods*, p. 370.

15. Eduardo Eurnekian and Baruch Tenembaum, 'Pastor Andre Trocme— the rescuer from Le Chambon-sur-Lignon', *The Jerusalem Post*, 5 April 2021, https://www.raoulwallenberg.net/General/Pastor-Andre-Trocme-The-Rescuer-From-Le-Chambon-Sur-Lignon/ (last accessed 8 September 2023).

16. Ian Sinclair, 'Resisting the Nazis in numerous ways: Nonviolence in occupied Europe, interview with George Paxton', Open Democracy, 19 July 2017, https://www.opendemocracy.net/en/non-violence-against-nazis-interview-with-george-paxton/ (last accessed 12 July 2023).

17. T. H. Tulchinsky, 'John Snow, cholera, the Broad Street Pump; waterborne diseases then and now', *Case Studies in Public Health*, 30 March 2018, pp. 77–99, https://www.ncbi.nlm.nih.gov/pmc/articles/PMC7150208/ (last accessed 13 July 2023).

18. See Paul Gregoire, 'Australia has a long history of fighting harm minimization drug policies', *Vice*, 11 January 2016, https://www.vice.com/en/article/4wbevq/how-australia-led-the-world-in-progressive-drug-policy-then-went-backwards (last accessed 13 July 2023); and Sarah

Larney, 'A global picture of injecting drug use, HIV and anti-HCV prevalence among people who inject drugs, and coverage of harm reduction interventions', Drug and Alcohol Research Connections, December 2017, http://connections.edu.au/researchfocus/global-picture-injecting-drug-use-hiv-and-anti-hcv-prevalence-among-people-who-inject (last accessed 13 July 2023).

19. C. K. Cassel, 'The Nevada desert demonstration', Medicine and War 3, no. 3 (1987), pp. 141–3, http://www.jstor.org/stable/45353120 (last accessed 13 July 2023).

20. General Medical Council, Good Medical Practice, 29 April 2019, https://www.gmc-uk.org/ethical-guidance/ethical-guidance-for-doctors/good-medical-practice/duties-of-a-doctor (last accessed 13 July 2023).

21. Office for Health Improvement & Disparities, 'Guidance: Climate and health: Applying "All Our Health"', 18 May 2022, https://www.gov.uk/government/publications/climate-change-applying-all-our-health/climate-and-health-applying-all-our-health (last accessed 13 July 2023).

22. Lancet Editorial, 'Doctors and civil disobedience', The Lancet 395 (25 January 2020), p. 248, https://doi.org/10.1016/S0140–6736(20)30120–3 (last accessed 13 July 2023).

23. Sydney Kamen, 'The world's first climate change conflict continues', Think Global Health, 10 December 2021, https://www.thinkglobalhealth.org/article/worlds-first-climate-change-conflict-continues (last accessed 13 July 2023).

24. Shannon Meade, 'Infrastructural inequalities in natural disasters: Hurricane Katrina case study through a critical infrastructure studies lens', 26 April 2021, https://storymaps.arcgis.com/stories/8bc180c2e72d47939e701abf17187e34 (last accessed 12 July 2023).

25. Juliette Landphair, 'The forgotten people of New Orleans': Community, vulnerability, and the Lower Ninth Ward, Journal of American History 94, no. 3 (December 2007), pp. 837–45, https://doi.org/10.2307/25095146 (last accessed 12 July 2023).

26. Lynne Jones, Outside the Asylum: A Memoir of War, Disaster and Humanitarian Psychiatry (London: Weidenfeld and Nicolson, 2017).

27. MedGlobal, 'Climate change, war, displacement, and health: The

impact on Syrian refugee camps', 20 September 2022, https://medg-lobal.org/climate-change-war-displacement-and-health-the-impact-on-syrian-refugee-camps/ (last accessed 13 July 2023).

28. Lynne Jones, *The Migrant Diaries* (New York: Refuge Press, 2021).

29. Colin Yeo, 'Briefing: The duty of refugee sea rescue in international law', Free Movement, 5 October 2021, https://freemovement.org.uk/refugee-sea-rescue-in-international-law-and-uk-law/ (last accessed 13 July 2023).

30. Daniel Howden, 'Europe's new anti-migrant strategy? Blame the rescuers', *Prospect* (20 March 2018), https://www.prospectmagazine.co.uk/magazine/europes-new-anti-migrant-strategy-blame-the-rescuers (last accessed 13 July 2023).

31. Iuventa: Solidarity at Sea, 'The case', https://iuventa-crew.org/en/case (last accessed 13 July 2023).

32. Nosheen Iqbal, 'Stansted 15: "We are not terrorists, no lives were at risk. We have no regrets"', *The Guardian*, 16 December 2018, https://www.theguardian.com/world/2018/dec/16/migrants-deportation-stansted-actvists (last accessed 13 July 2023).

33. Ben Smoke, 'The Stansted 15's quashed conviction shows we were never terrorists', *The Guardian*, 2 February 2021, https://www.theguardian.com/commentisfree/2021/feb/02/stansted-15-quashed-conviction-terrorists-deportation-hostile-environment (last accessed 13 July 2023).

34. Damien Gayle, 'Stansted 15: No jail for activists convicted of terror-related offences', *The Guardian*, 6 February 2019, https://www.theguardian.com/global/2019/feb/06/stansted-15-rights-campaigners-urge-judge-to-show-leniency (last accessed 13 July 2023).

35. Smoke, 'The Stansted 15's quashed conviction shows we were never terrorists'.

36. David Barnett, 'The absurdity of owning moors and mountains', *The Independent*, 27 April 2022, https://www.independent.co.uk/indepen-dentpremium/long-reads/owner-moors-mountains-kinder-scout-mass-trespass-b2063359.html (last accessed 14 July 2023).

37. https://www.hs2rebellion.earth/ (last accessed 14 July 2023).

38. 'How Greenpeace creates change', Greenpeace, https://www.green-peace.org.uk/about-greenpeace/how-we-create-change/ (last accessed 14 July 2023).

39. 'How Greenpeace changed the world', Greenpeace, https://www.green-peace.org.uk/about-greenpeace/victories/ (last accessed 14 July 2023).

40. Jasmine Watkiss, 'Greenpeace UK creates underwater boulder barrier in the South West Deeps to block destructive industrial fishing', Greenpeace, 2 September 2022, https://www.greenpeace.org.uk/news/breaking-greenpeace-uk-creates-underwater-boulder-barrier-in-the-south-west-deeps-to-block-destructive-industrial-fishing/ (last accessed 14 July 2023).

41. Adam Bychawski, 'Just Stop Oil didn't delay us getting to M20 crash, says ambulance service', Open Democracy, 25 October 2022, https://www.opendemocracy.net/en/just-stop-oil-dartford-ambulance-not-delayed-m20-accident/ (last accessed 14 July 2023).

42. Author interview with Patrick Hart, Zoom, 26 February 2023.

43. 'Open letter to the UK judiciary: The courts are now a site of nonviolent civil resistance', Insulate Britain, 20 April 2022, https://insulatebritain.com/2022/04/20/open-letter-to-the-uk-judiciarythe-courts-are-now-a-site-of-nonviolent-civil-resistance/ (last accessed 22 July 2023).

44. 'Today in legal history: Six Committee of 100 activists go on trial, for breaching Official Secrets Act, 1962', London Radical Histories, 12 February 2017, https://pasttense.co.uk/2017/02/12/today-in-legal-history-six-committee-of-100-activists-go-on-trial-for-breaching-official-secrets-act-1962/ (last accessed 22 July 2023). Quotation from Richard Taylor, *Against the Bomb: The British Peace Movement, 1958–1965* (Oxford: Oxford University Press, 1995).

45. Levy, *Ban the Bomb!*, pp. 134–9.

46. Michael Randle, *Rebel Verdict* (Sparsnäs: Irene, 2022).

47. Ibid., pp. 459–61.

48. 'Just Stop Oil supporters who resisted the criminality of Exxon Mobil, acquitted of wrongdoing', Just Stop Oil, 25 January 2023, https://just-stopoil.org/2023/01/25/just-stop-oil-supporters-who-resisted-the-criminality-of-exxon-mobil-acquitted-of-wrongdoing/ (last accessed 14 July 2023).

49. Damien Gayle, 'Insulate Britain activist jailed for eight weeks for contempt of court', *The Guardian*, 7 February 2023. Nixon was sentenced

to an additional five weeks for the road blocking in April 2023. See Anita Mureithi, 'Insulate Britain activists jailed after telling judge: "We won't stop"', Open Democracy, 21 April 2023, https://www.opendemocracy.net/en/insulate-britain-civil-resistance-protests-climate-crisis (last accessed 14 July 2023).

50. Anita Murethi, 'Activists jailed for seven weeks for defying ban on mentioning climate crisis', Open Democracy, 3 March 2023, https://www.opendemocracy.net/en/activists-jailed-for-seven-weeks-for-defying-ban-on-mentioning-climate-crisis/ (last accessed 14 July 2023).

51. 'Top UK lawyers attend Insulate Britain contempt hearing at Inner London Crown Court', Insulate Britain, 3 March 2023, Video, https://www.youtube.com/watch?v=hTm8eIzza84 (last accessed 14 July 2023).

52. '"You should feel guilty for nothing" says judge, as he finds seven guilty and acquits two, for disrupting Esso terminal in Birmingham', Just Stop Oil, 16 February 2023, https://juststopoil.org/2023/02/16/you-should-feel-guilty-for-nothing-says-judge-as-he-finds-seven-guilty-and-aquits-two-for-disrupting-oil-supplies-at-esso-terminal-in-birmingham/ (last accessed 14 July 2023).

53. Jane Atkinson [Lynne Jones], 'The woman behind Solidarity: The story of Anna Walentynowicz', *Ms Magazine* 12, no. 8 (February 1984), pp. 96–8.

7. WE CAN JUST SAY NO

1. Sharp, *Politics of Nonviolent Action, Part Two: The Methods*, p. 184.

2. Thompson, *Protest and Survive*, p. 31.

3. Rachel Adams [Lynne Jones], 'Dialogue across the Iron Curtain', *New Statesman*, 12 October 1984, pp. 21–2.

4. Neal Ascherson, *The Struggles for Poland* (New York: Random House, 1987), p. 223.

5. Ibid., p. 215.

6. Norman Davies, *Heart of Europe: A Short History of Poland* (Oxford: Oxford University Press, 1986), p. 388.

7. See WiP Letter [WP6], in Lynne Mastnak [Lynne Jones], 'The process of engagement in non-violent collective action: Case studies from the 1980s', unpublished PhD thesis, Bath University, 1995, p. 221; and

R. Adams [Lynne Jones], 'Take not that hypocritic oath', *New Statesman*, 4 April 1986, p. 19.

8. Mastnak, 'The process of engagement'.

9. WiP Letter to END Convention in Amsterdam, July 1985 [WP2], in Mastnak, 'The process of engagement', p. 215. Father Jerzy Popiełuszko was a Polish Catholic priest associated with Solidarity who was murdered by three security agents from the Polish Ministry of the Interior in 1984.

10. Jarosław Dubiel, 'The spirit of Otto Schimek', Voice of Solidarity, April 1986, in Mastnak, 'The process of engagement', p. 220.

11. Mastnak, 'The process of engagement'.

12. Zein Nakhoda, 'Solidarność (Solidarity) brings down the Communist government of Poland, 1988–89', Global Nonviolent Action Database, 10 September 2011, https://nvdatabase.swarthmore.edu/content/solidarno-solidarity-brings-down-communist-government-poland-1988-89 (last accessed 14 July 2023).

13. Henry David Thoreau, 'Civil disobedience (1849)', in Bob Blaisdell (ed.), *Essays on Civil Disobedience* (New York: Dover Publications, 2016), pp. 22–42.

14. Henry David Thoreau, 'Slavery in Massachusetts (1854)', in Blaisdell (ed.), *Essays on Civil Disobedience*, pp. 43–56.

15. Mahatma Gandhi, 'Satyagraha (Noncooperation) (1920)', in Blaisdell (ed.), *Essays on Civil Disobedience*, pp. 92–4.

16. Martin Luther King Jr, 'Love, law, and civil disobedience (1961)', in Blaisdell, ed. *Essays on Civil Disobedience*, pp. 126–37.

17. Elizabeth Heyrick, *Immediate, Not Gradual Abolition, or, An Inquiry into the Shortest, Safest, and most Effectual Means of Getting Rid of West Indian Slavery* (Boston, MA: Isaac Knapp, 1838), available at Library of Congress: African American Pamphlet Collection copy, pp. 4, 8, 34, https://www.loc.gov/item/11009325/ (last accessed 14 July 2023).

18. John Simkin, 'Women and the anti-slavery movement', Spartacus Educational, September 1997, updated January 2020, https://spartacus-educational.com/REslaveryW.htm (last accessed 16 July 2023).

19. Chenoweth, *Civil Resistance*, p. 98; Sharp, *Politics of Nonviolent Action, Part Two: The Methods*, p. 227.

20. Sharp, *Politics of Nonviolent Action, Part Two: The Methods*, p. 189.
21. Gene Sharp, *Tyranny Could Not Quell Them! How Norway's Teachers Defeated Quisling during the Nazi Occupation and What it Means for Unarmed Defence Today*, Peace News Pamphlet, London, 1959, and https://onlinebooks.library.upenn.edu/webbin/book/lookupid?key=olbp73258 (last accessed 16 July 2023).
22. Sharp, *Politics of Nonviolent Action, Part Two: The Methods*, p. 224.
23. 'Birmingham Campaign, April 3, 1963 to May 10, 1963', Stanford University, The Martin Luther King, Jr. Research and Education Institute, https://kinginstitute.stanford.edu/encyclopedia/birmingham-campaign (last accessed 1 June 2023).
24. Destin Jenkins, 'What history can teach banks about making change', Deal Book Newsletter, *The New York Times*, 1 May 2021, https://www.nytimes.com/2021/05/01/business/dealbook/history-banks-social-movements.html (last accessed 16 July 2023).
25. 'Workers united: The Delano grape strike and boycott', National Park Service, César E. Chávez National Monument, 19 December 2022, https://www.nps.gov/articles/000/workers-united-the-delano-grape-strike-and-boycott.htm (last accessed 16 July 2023).
26. Sharp, *Politics of Nonviolent Action, Part Two: The Methods*, p. 317.
27. Ibid., p. 329.
28. Randle, *Civil Resistance*, p. 195.
29. '1990: One in five yet to pay poll tax', BBC, 'On This Day', 14 August 1990, http://news.bbc.co.uk/onthisday/hi/dates/stories/august/14/newsid_2495000/2495911.stm (last accessed 16 July 2023).
30. 'Crisis year 2022 brought $134 billion in excess profit to the West's five largest oil and gas companies', Global Witness, 9 February 2023, https://www.globalwitness.org/en/campaigns/fossil-gas/crisis-year-2022-brought-134-billion-in-excess-profit-to-the-wests-five-largest-oil-and-gas-companies/ (last accessed 16 July 2023).
31. 'Strike dates: Who is striking and what pay do they want?', BBC News, 7 July 2023, https://www.bbc.co.uk/news/business-62134314 (last accessed 16 July 2023).
32. Philippa Nuttall, '"It was my moral duty": Why a Swiss father went on hunger strike for the climate', *New Statesman*, 14 December 2021,

https://www.newstatesman.com/environment/climate/2021/12/it-was-my-moral-duty-why-a-swiss-father-went-on-hunger-strike-for-the-climate (last accessed 16 July 2023).

33. Author interview with Angus Rose, Westminster, London, 12 April 2022.

34. 'Senior health professionals call for urgent climate briefing of all MPs by the chief scientific adviser: Open letter to PM', *BMJ* 377, 16 April 2022, https://www.bmj.com/content/377/bmj.o987 (last accessed 16 July 2023).

35. 'Sir Patrick Vallance briefs parliamentarians on climate change', Policy Connect: Parliamentary Briefings, 12 July 2022, https://www.policy-connect.org.uk/news/sir-patrick-vallance-briefs-parliamentarians-cli-mate-change

36. Damian Carrington, 'Desmond Tutu calls for anti-apartheid style boy-cott of fossil fuel industry', *The Guardian*, 10 April 2014, https://www.theguardian.com/environment/2014/apr/10/desmond-tutu-anti-apart-heid-style-boycott-fossil-fuel-industry (last accessed 16 July 2023).

37. 'Boycott South African goods', Forward to Freedom: The History of the British Anti-Apartheid Movement, 1959–1994, https://www.aamarchives.org/campaigns/boycott.html (last accessed 16 July 2023).

38. Ryle Dwyer, 'State archives: Dunnes Stores strike demonstrated power of the few', *The Irish Examiner*, 1 January 2016, https://www.irishex-aminer.com/news/arid-20373917.html (last accessed 16 July 2023).

39. Ryan Leitner, 'British students force end of Barclays Bank's invest-ments in South African Apartheid 1969–1987', Global Nonviolent Action Database, 6 February 2014, https://nvdatabase.swarthmore.edu/content/british-students-force-end-barclays-bank-s-investments-south-african-apartheid-1969–1987 (last accessed 16 July 2023).

40. Lori Campbell, 'Revealed: Britain's worst banks for emissions (and the best)', Good with Money, 12 October 2022, https://good-with-money.com/2022/10/12/revealed-britains-worst-banks-for-emissions-and-the-best/ (last accessed 12 July 2023).

41. Global Commitments Divestment Database, https://divestmentdata-base.org/ (last accessed 16 July 2023).

42. David Carlin, 'The case for fossil fuel divestment', Forbes, 20 February

2021, https://www.forbes.com/sites/davidcarlin/2021/02/20/the-case-for-fossil-fuel-divestment (last accessed 29 August 2023).

43. Mattha Busby, 'Capitalism is part of solution to climate crisis, says Mark Carney', *The Guardian*, 31 July 2019, https://www.theguardian.com/business/2019/jul/31/capitalism-is-part-of-solution-to-climate-crisis-says-mark-carney (last accessed 16 July 2023).

44. Harriet Reuter Hapgood and Peter Bosshard, '2022 scorecard on insurance, fossil fuels and the climate emergency', Insure Our Future, 19 October 2022, https://reclaimfinance.org/site/en/2022/10/19/2022-scorecard-on-insurance-fossil-fuels-and-the-climate-emergence-insure-our-future/ (last accessed 16 July 2023).

45. 'Ugandan students arrested by local police during protests against EACOP Project', #StopEACOP, 4 October 2022, https://www.stopeacop.net/our-news/breaking-ugandan-students-arrested-by-local-police-during-protests-against-eacop-project (last accessed 16 July 2023).

46. Gerard Kreeft, 'Will the East African Crude Oil Pipeline (EACOP) ever be built?' 17 November 2022, Institute for Energy Economics and Financial Analysis. https://ieefa.org/resources/will-east-african-crude-oil-pipeline-eacop-ever-be-built (last accessed 23 October 2023).

47. 'The British Museum and BP—will they or won't they?', Real Media, 1 March 2023, https://realmedia.press/british-museum-will-they-wont-they/ (last accessed 16 July 2023).

48. 'Science Museum cancels event as speakers withdraw over fossil fuel sponsorship', Culture Unstained, 27 April 2022, https://cultureunstained.org/2022/04/27/science-museum-cancels-event-as-speakers-withdraw-over-fossil-fuel-sponsorship/ (last accessed 16 July 2023).

49. Ellie Harrison, 'Royal Opera House ends sponsorship relationship with BP after 33 years', *The Independent*, 25 January 2023, https://www.independent.co.uk/arts-entertainment/music/news/royal-opera-house-bp-sponsorship-b2268992.html (last accessed 16 July 2023).

50. Mark Rylance, 'Why I'm resigning from the RSC', Culture Unstained, 21 June 2019, https://cultureunstained.org/2019/06/21/mark-rylance-rsc/ (last accessed 16 July 2023).

51. Harriet Whitehead, 'Young people threaten to boycott Royal Shakespeare

Company over BP deal', Civil Society, 26 September 2019, https://www.civilsociety.co.uk/news/young-people-threaten-to-boycott-royal-shakespeare-company-over-oil-giant-deal.html (last accessed 16 July 2023).

52. Matthew Taylor, 'Royal Shakespeare Company to end BP sponsorship deal', *The Guardian*, 2 October 2019, https://www.theguardian.com/stage/2019/oct/02/royal-shakespeare-company-to-end-bp-sponsorship-deal https://fossilfreeresearch.org/letter/ (last accessed 16 July 2023).

53. 'Our letter', Fossil Free Research, https://fossilfreeresearch.org/letter/ (last accessed 16 July 2023).

54. Princeton University, 'Fossil fuel dissociation', https://fossilfueldissociation.princeton.edu/background (last accessed 6 September 2023).

55. Waseem Mohamed and Damian Carrington, 'Fossil fuel recruiters banned from UK university careers service', *The Guardian*, 28 September 2022, https://www.theguardian.com/environment/2022/sep/28/fossil-fuel-recruiters-banned-from-birkbeck-university-if-london-careers-service (last accessed 16 July 2023).

56. Alex Lawson, 'Consultant who ditched Shell: "Take a look at yourselves in the mirror"', *The Guardian*, 28 May 2022, https://www.theguardian.com/business/2022/may/28/consultant-who-ditched-shell-take-a-look-at-yourselves-in-the-mirror (last accessed 16 July 2023).

57. 'AR6 Synthesis Report: Headline statements', IPCC, 2023, https://www.ipcc.ch/report/ar6/syr/resources/spm-headline-statements/ (last accessed 16 July 2023).

58. Author interview with Melinda Janki, London, 23 February 2023.

59. 'Take the jump launches research: The power of people', Take the Jump, 7 March 2023, https://takethejump.org/latest/jump-launches-research-the-power-of-people (last accessed 16 July 2023).

60. Gee Harland, 'Fossil fuel campaigner refuses to pay council tax despite order', *Bucks Free Press*, 23 March 2023, https://www.bucksfreepress.co.uk/news/23407399.fossil-fuel-campaigner-refuses-pay-council-tax-despite-order/ (last accessed 16 July 2023).

61. Mark Jones, 'Gary Lineker details tearful reaction to BBC boycott from Ian Wright and Alan Shearer', *The Daily Mirror*, 27 March 2023,

https://www.mirror.co.uk/sport/football/news/gary-lineker-bbc-wright-shearer-29561944 (last accessed 16 July 2023).

62. 'Declaration of conscience', Lawyers Are Responsible, https://www.lar.earth/sign/ (last accessed 16 July 2023).

63. LawyersAreResponsible (@LawyersAreResp), Twitter, 29 March 2023, 9:48 AM, https://twitter.com/LawyersAreResp/status/164099947501 4516736 (last accessed 16 July 2023).

64. Thoreau, 'Civil Disobedience (1849)', p. 22.

65. Colin Fernandez, Alice Wright and Kumail Jaffer, 'Fury at woke barristers refusing to prosecute eco warriors', *Mail Online*, 23 March 2023, https://www.dailymail.co.uk/news/article-11896725/Fury-woke-barristers-refusing-prosecute-eco-warriors.html (last accessed 16 July 2023).

66. Author interview with Tim Crosland, London, 27 January 2023.

67. Jolyon Maugham, 'Why I'm joining more than 100 lawyers in refusing to prosecute climate protesters', *The Guardian*, 24 March 2023, https://www.theguardian.com/commentisfree/2023/mar/24/100-lawyers-prosecute-climate-protesters-laws-planet-criminalise (last accessed 16 July 2023).

8. EVERYTHING IS CONNECTED

1. PBI Guatemala, 'Mandate and principles', https://pbi-guatemala.org/en/about-pbi-guatemala/mandate-and-principles (last accessed 19 July 2023).

2. PBI Guatemala, Monthly Information Package 229, October 2022, https://pbi-guatemala.org/en/multimedia/monthly-information-packages (last accessed 19 July 2023).

3. Alba Kapoor, Nannette Youssef and Simon Hood, 'Confronting injustice: Racism and the environmental emergency', Greenpeace UK and Runnymede Trust, 2022, https://www.greenpeace.org.uk/challenges/environmental-justice/race-environmental-emergency-report/ (last accessed 20 July 2023).

4. 'How Indigenous peoples enrich climate action', United Nations: Climate Change, 9 August 2022, https://unfccc.int/news/how-indigenous-peoples-enrich-climate-action (last accessed 20 July 2023). The UNFCCC

Secretariat (UN Climate Change) is the United Nations entity tasked with supporting the global response to the threat of climate change. UNFCCC stands for United Nations Framework Convention on Climate Change. The convention has near universal membership (198 Parties) and is the parent treaty of the 2015 Paris Agreement.

5. 'Census results for the Republic of Guatemala, 2018', Resultados del Censo 2018, https://www.censopoblacion.gt/mapas (last accessed 20 July 2023).

6. Guatemala Commission for Historical Clarification (CEH), *Guatemala, Memory of Silence: Report of the Commission for Historical Clarification*, CEH, 1999, pp. 17–23.

7. 'Guatemala', Center for Justice & Accountability, https://cja.org/where-we-work/guatemala/ (last accessed 20 July 2023).

8. Aurora Muñoz, 'Guatemalans overthrow a dictator, 1944', Global Nonviolent Action Database, 30 November 2009, https://nvdatabase.swarthmore.edu/content/guatemalans-overthrow-dictator-1944 (last accessed 20 July 2023).

9. Jeff Abbott, 'Guatemala: Former president sentenced to 16 years for corruption', *The Guardian*, 8 December 2022, https://www.theguardian.com/world/2022/dec/08/guatemala-president-otto-perez-molina-corruption-sentence (last accessed 20 July 2023).

10. Nina Lakhani, 'Guatemalan president's downfall marks success for corruption investigators', *The Guardian*, 9 September 2015, https://www.theguardian.com/world/2015/sep/09/guatemala-president-otto-perez-molina-cicig-corruption-investigation (last accessed 20 July 2023).

11. 'Fernando García case', National Security Archive, 2013, https://nsarchive.gwu.edu/fernando-garcia-case (last accessed 20 July 2023).

12. Author interview with Carlos Juárez, Guatemala City, Guatemala, 15 October 2022.

13. See Ana Castro, 'Guatemala's anti-corruption commission ordered to close by President Morales', Global Initiative against Transnational Crime, 17 September 2019, https://globalinitiative.net/analysis/guatemalas-anti-corruption-commission-ordered-to-close-by-president-morales/ (last accessed 20 July 2023); and Alex Papadovassilakis and Shane Sullivan, 'Guatemala's war between Morales and CICIG not over

yet', 22 April 2021, InSight Crime, https://insightcrime.org/news/cicig-guatemala-president-morales/ (last accessed 20 July 2023).

14. 'Guatemala: Events of 2022', in World Report 2023, Human Rights Watch, https://www.hrw.org/world-report/2023/country-chapters/guatemala (last accessed 20 July 2023).

15. Ibid.

16. 'Guatemala: Justice under attack: Amnesty International Report to the Working Group of the 42nd Session of the UN Universal Periodic Review, January 2023', Amnesty International, 30 June 1922, https://www.amnesty.org/en/documents/amr34/5738/2022/en/ (last accessed 20 July 2023).

17. 'Trial of *El Periódico* founder Jose Rubén Zamora set to begin on December 8', Media Development Investment Fund, 29 November 2022, https://www.mdif.org/trial-el-periodico-editor-jose-ruben-zamora-begin-december/ (last accessed 20 July 2023).

18. Author interview with Carlos Juárez, Guatemala City, Guatemala, 15 October 2022.

19. Author interview with Silvia Weber and Laura Gomariz, Guatemala City, Guatemala, 15 October 2022.

20. Interview with Bernardo Caal Xol by Silvia Weber, PBI Guatemala, Acércate, 25 March 2021, https://www.youtube.com/watch?v=GaB_24iH2OM (last accessed 20 July 2023).

21. Author interview with Bernardo Caal Xol, Antigua, Guatemala, 19 October 2022.

22. 'Guatemala ratifies convention guaranteeing Indigenous rights', International Labour Organization, 13 June 1996, https://www.ilo.org/global/about-the-ilo/newsroom/news/WCMS_008061/lang—en/index.htm (last accessed 20 July 2023).

23. 'OXEC S.A. lawsuit (re consultation for hydroelectric plants, Guatemala)', Business and Human Rights Resource Centre, 2017, https://www.business-humanrights.org/en/latest-news/oxec-sa-lawsuit-re-consultation-for-hydroelectric-plants-guatemala/ (last accessed 20 July 2023).

24. Author interview with Bernardo Caal Xol, Antigua, Guatemala, 19 October 2022.

25. Author interview with José Bo and his family, Cahabón, Guatemala, 21 October 2022.

26. Sara Zaske, 'Methane emissions from reservoirs are increasing', WSU Insider, 19 September 2022, https://news.wsu.edu/news/2022/09/19/methane-emissions-from-reservoirs-are-increasing/(last accessed 23 October 2023).

27. 'An appeal to the UNFCCC & State parties at COP27: Put human rights at the centre of climate action', Indigenous People's Rights International, 12 October 2022, https://iprights.org/index.php/en/all-news/an-appeal-to-the-unfccc-state-parties-at-cop27 (last accessed 20 July 2023).

28. 'When jail doesn't stop those who defend the land: The case of Bernardo Caal', interview with Bernardo Caal by International Land Coalition, 15 July 2022, https://lac.landcoalition.org/en/noticias/cuando-la-carcel-no-detiene-a-quienes-defienden-la-tierra-el-caso-de-bernardo-caal/ (last accessed 20 July 2023).

29. Author interview with Don Mario Chocoj, Cahabón, Guatemala, 23 October 2022.

30. Author interview with Paolino Temo, Cahabón, Guatemala, 24 October 2022.

31. Author interview with Maria Caal Xol, Cahabón, Guatemala, 22 October 2022.

32. Author interview with Don Mario Chocoj, Cahabón, Guatemala, 23 October 2022.

33. Madeleine Cuff, 'Water companies are dumping raw sewage into rivers and seas without permits, Dispatches investigation reveals', inews, 29 August 2022, https://inews.co.uk/news/water-companies-dumping-raw-sewage-rivers-seas-without-permits-dispatches-investigation-reveals-1820683 (last accessed 3 August 2023).

34. Museo Nacional de Arte de Guatemala (MUNAG), http://cultura-guate.com/munag/ (last accessed 9 September 2023).

35. See https://www.facebook.com/majawil.qij.9/ (last accessed 3 August 2023).

36. Mastnak, 'The process of engagement', p. 400.

37. Ana María Méndez Dardón, 'Regressive wave for women in Central

America', WOLA, 8 March 2023, ps://www.wola.org/analysis/regressive-wave-women-central-america/ (last accessed 20 July 2023).

38. Author interview with María Morales, Chichicastenango, Guatemala, November 2022.

39. Press release, Abya Yala, 12 October 2022, III Cumbre de Mujeres Indígenas de Abya Yala ps://www.facebook.com/photo/?fbid=136580 042459207&set=pcb.136580932459118 (last accessed 20 July 2023).

40. Jenny Harper Gow, 'Women's resistance speech', 12 October 2019, Russell Square, London (text sent to author 18 October 2019).

41. Manuel Chavajay, 'La Performance', No-Ficción (@noficciongt), Twitter, 12 November 2022, 1:36 PM, https://twitter.com/noficciongt/status/1591424608246398978?s=20 (last accessed 20 July 2023).

42. Nancy González, 'El plástico, violencia instaurada, silenciosa y dispersa en el Lago Atitlán', Prensa Comunitaria, 24 October 2022, https://prensacomunitaria.org/2022/10/el-plastico-violencia-instaurada-silenciosa-y-dispersa-en-el-lago-atitlan/ (last accessed 20 July 2023).

43. Interview with Nancy Gonzalez, San Pedro de la Laguna, Guatemala, 4th November 2022.

44. 'The grandmother lake: Conservation and colonialism in Guatemala', Al Jazeera English, 23 April 2021, https://interactive.aljazeera.com/aje/2021/guatemala-conservation-colonialism/index.html (last accessed 20 July 2023).

45. Interview with Nancy Gonzalez, San Pedro de la Laguna, Guatemala, 4th November 2022.

46. 'The grandmother lake'.

47. Rebecca Lindsey and LuAnn Dahlman, 'Climate change: Global temperature', NOAA Climate.gov, 18 January 2023, https://www.climate.gov/news-features/understanding-climate/climate-change-global-temperature (last accessed 16 July 2023); Rebecca Lindsey, 'Climate change: Atmospheric carbon dioxide', 12 May 2023, https://www.climate.gov/news-features/understanding-climate/climate-change-atmospheric-carbon-dioxide; and NASA Global climate change: Vital signs of the planet, 'Carbon dioxide', https://climate.nasa.gov/vital-signs/carbon-dioxide/ (last accessed 14 September 2023).

48. 'Horn of Africa drought regional humanitarian overview & call to action

(revised 26 May 2023)', Reliefweb, OCHA, https://reliefweb.int/report/ethiopia/horn-africa-drought-regional-humanitarian-overview-call-action-revised-26-may-2023 (last accessed 16 July 2023).

49. Oxfam International, 'One person likely dying from hunger every 48 seconds in drought-ravaged East Africa as world again fails to heed warnings', 17 May 2022, https://www.oxfam.org/en/press-releases/one-person-likely-dying-hunger-every-48-seconds-drought-ravaged-east-africa-world (last accessed 16 July 2023).

50. Tim Gore, 'Confronting carbon inequality', Oxfam International, 21 September 2020, https://www.oxfam.org/en/research/confronting-carbon-inequality (last accessed 16 July 2023).

51. Niko Kommenda, 'How your flight emits as much CO_2 as many people do in a year', *The Guardian*, 19 July 2019, https://www.theguardian.com/environment/ng-interactive/2019/jul/19/carbon-calculator-how-taking-one-flight-emits-as-much-as-many-people-do-in-a-year (last accessed 16 July 2023).

52. Robert Booth, 'Extinction Rebellion announces move away from disruptive tactics', *The Guardian*, 1 January 2023, https://www.theguardian.com/world/2023/jan/01/extinction-rebellion-announces-move-away-from-disruptive-tactics (last accessed 1 September 2023).

53. Madeleine Cuff, 'How can we keep homes cool in extreme heat without air conditioning?', *New Scientist*, 13 July 2023, https://www.newscientist.com/article/2382452-how-can-we-keep-homes-cool-in-extreme-heat-without-air-conditioning/ (last accessed 16 July 2023).

54. Greg Barradale, 'Not a single arrest made during Extinction Rebellion's four-day "The Big One" protest', *The Big Issue*, 25 April 2023, https://www.bigissue.com/news/activism/not-a-single-arrest-made-during-extinction-rebellions-four-day-the-big-one-protest/ (last accessed 16 July 2023).

55. Interview: 'Hugh Brasher London Marathon race director', Extinction Rebellion UK, 23 April 2023, https://www.youtube.com/watch?v=pRNdd20NYJ4 (last accessed 16 July 2023).

56. 'Public Order Act: New protest offences & "serious disruption"', Liberty, 3 July 2023, https://www.libertyhumanrights.org.uk/advice_information/public-order-act-new-protest-offences/ (last accessed 16 July 2023).

57. Home Office (@ukhomeoffice), Twitter, 26 April 2023, 4:25 PM, https://twitter.com/ukhomeoffice/status/1651246232361156608?s=20 (last accessed 10 September 2023).

58. 'UN human rights chief urges UK to reverse "deeply troubling" Public Order Bill', United Nations Human Rights, Office of the Commissioner, 27 April 2023, ps://www.ohchr.org/en/press-releases/2023/04/un-human-rights-chief-urges-uk-reverse-deeply-troubling-public-order-bill (last accessed 16 July 2023).

59. Helena Horton and agencies, 'Just Stop Oil protesters jailed for Dartford Crossing protest', *The Guardian*, 21 April 2023, https://www.the-guardian.com/environment/2023/apr/21/just-stop-oil-protesters-jailed-for-dartford-crossing-protest (last accessed 16 July 2023).

60. 'King Charles III coronation: King Charles sits in the coronation chair and gets jewelled sword', WION, 6 May 2023, https://www.youtube.com/watch?v=0FTpS5wW6q4 (last accessed 16 July 2023).

61. See 'Just Stop Oil forcibly removed from the Home Affairs Committee whilst eight arrested outside parliament', Just Stop Oil, 17 May 2023, https://juststopoil.org/2023/05/17/just-stop-oil-forcibly-removed-from-the-home-affairs-committee-whilst-eight-arrested-outside-par-liament/ (last accessed 16 July 2023); and Lizzie Dearden, 'Met Police deny political pressure influenced coronation protest arrests', *The Independent*, 17 May 2023, https://www.independent.co.uk/news/uk/home-news/just-stop-oil-coronation-met-police-arrests-b2340465.html (last accessed 16 July 2023).

62. Denis Campbell, 'London NHS trust cancels operations as IT system fails in heatwave', *The Guardian*, 21 July 2022, https://www.theguard-ian.com/society/2022/jul/21/london-nhs-trust-cancels-operations-as-it-system-fails-in-heatwave (last accessed 16 July 2023).

63. Office for National Statistics and UK Health Security Agency, 'Excess mortality during heat-periods: 1 June to 31 August 2022', Office for National Statistics, 7 October 2022, ps://www.ons.gov.uk/peoplepop-ulationandcommunity/birthsdeathsandmarriages/deaths/articles/excessmortalityduringheatperiods/englandandwales1juneto31august2022#main-points (last accessed 16 July 2023).

64. Joan Ballester et al., 'Heat-related mortality in Europe during the sum-

mer of 2022', *Nature Medicine* 29 (July 2023), https://www.nature.com/articles/s41591-023-02419-z (last accessed 16 July 2023).

65. Alexandre Tuel and Elfatih A. B. Eltahir, 'This summer has been a glimpse into Europe's hot, dry future', *Bulletin of the Atomic Scientists*, 5 September 2022, https://thebulletin.org/2022/09/this-summer-has-been-a-glimpse-into-europes-hot-dry-future (last accessed 16 July 2023).

66. Denise Chow, 'Deadly combo of extreme heat and air pollution grips parts of Southeast Asia', NBC News, 10 May 2023, https://www.nbc-news.com/science/environment/deadly-combo-extreme-heat-air-pollution-grips-parts-southeast-asia-rcna83419 (last accessed 16 July 2023).

67. Leyland Cecco, 'Canada: Extreme "heat dome" temperatures set to worsen wildfires', *The Guardian*, 12 May 2023, https://www.theguardian.com/world/2023/may/12/western-canada-record-temperatures-heat-dome-wildfires (last accessed 16 July 2023).

68. Damian Carrington, 'Ecosystem collapse "inevitable" unless wildlife losses reversed', *The Guardian*, 24 February 2023, https://www.theguardian.com/environment/2023/feb/24/ecosystem-collapse-wildlife-losses-permian-triassic-mass-extinction-study?CMP=Share_AndroidApp_Othertinction (last accessed 16 July 2023).

69. 'How climate change is damaging the oceans', *Zurich*, 8 June 2023, https://www.zurich.com/en/media/magazine/2022/the-relationship-between-our-oceans-and-climate-change (last accessed 16 July 2023).

70. Global Commission on the Economics of Water, 'Turning the tide: A call to collective action', March 2023, https://turningthetide.water-commission.org/ (last accessed 16 July 2023).

71. Todd Miller, Nick Buxton and Mark Akkerman, 'Global climate wall: How the world's wealthiest nations prioritise borders over climate action', Transnational Institute, 25 October 2021, https://www.tni.org/en/publication/global-climate-wall (last accessed 4 March 2023).

72. Saeid Vaghefi et al., 'ChatClimate: Grounding conversational AI in climate science', 11 April 2023, available at SSRN, https://ssrn.com/abstract=4414628 (last accessed 31 May 2023).

73. 'ChatClimate—grounded on the latest IPCC report', ChatClimate, https://www.chatclimate.ai/ (answer accessed 31 May 2023).

9. LIFEBOATS

1. '1981: Penlee lifeboat disaster', RNLI Lifeboats, https://rnli.org/about-us/our-history/timeline/1981-penlee-lifeboat-disaster (last accessed 7 November 2023).

2. Sinead Butler, 'RNLI had a record year after Farage attacks and people are thrilled', *Indy100*, 2 January 2022, https://www.indy100.com/news/rnli-record-year-donations-nigel-farage-b1985559 (last accessed 16 July 2023).

3. 'Statement on the humanitarian work of the RNLI in the English Channel', RNLI Lifeboats, 28 July 2021, https://rnli.org/news-and-media/2021/july/28/statement-on-the-humanitarian-work-of-the-rnli-in-the-english-channel (last accessed 16 July 2023).

4. Katy Fallon, 'The Greek refugees battling to prevent Covid-19 with handmade face masks', *The Guardian*, 18 March 2020, https://www.theguardian.com/global-development/2020/mar/18/the-greek-refugees-battling-to-prevent-covid-19-with-handmade-face-masks (last accessed 16 July 2023).

5. See 'Greenham Common—A chronology', *The Guardian*, 5 September 2006, https://www.theguardian.com/uk/2006/sep/05/greenham4 (last accessed 16 July 2023); Jill Raymond, 'Still carrying Greenham home', *The Land* 29 (2021), https://www.thelandmagazine.org.uk/articles/still-carrying-greenham-home (last accessed 16 July 2023); and Sarah Hipperson, *Greenham: Non-Violent Women v The Crown Prerogative* (London: Greenham Publications, 2005).

6. Historic England, 'Cruise missile shelter complex, Greenham Common Airbase', https://historicengland.org.uk/listing/the-list/list-entry/1021040 (last accessed 8 September 2023).

7. From my own notes taken on 4 September 2021, and Sarah Bosley, 'Peace women return to Greenham Common to commemorate 40th anniversary of the Greenham Common Peace Camp', *Newburytoday*, 9 September 2021, https://www.newburytoday.co.uk/news/peace-camp-remembered-9215212/ (last accessed 16 July 2023).

8. The Pugwash Conferences on Science and World Affairs is an international network founded in 1957 that brings together scientists and others to work towards the elimination of weapons of mass destruction by

contributing to scientific, evidence-based governmental policymaking and by promoting international dialogue across divides. See British Pugwash, https://britishpugwash.org/what-is-pugwash/ (last accessed 16 July 2023).

9. Author interview with Rebecca Johnson, Zoom, 30 March 2023.

10. Hipperson, *Greenham*, p. 39.

11. 'What happens if nuclear weapons are used?', ICAN, https://www. icanw.org/catastrophic_harm (last accessed 16 July 2023).

12. R. Johnson, 'The Nuclear Ban and Humanitarian Strategies to Eliminate Nuclear Threats' in Nikolas Vik Steen and Olav Njølstad (eds.), *Nuclear Disarmament: A Critical Assessment* (Abingdon: Routledge, 2019), pp. 75–93.

13. Author interview with Rebecca Johnson, Zoom, 30 March 2023.

14. 'International Campaign to Abolish Nuclear Weapons (ICAN)-Facts', The Nobel Peace Prize, 2017, https://www.nobelprize.org/prizes/ peace/2017/ican/facts/ (last accessed 16 July 2023).

15. Rebecca Eleanor Johnson, 'Nuclear weapons are banned! What does this mean for Britain?', Acronym Institute for Disarmament Diplomacy (AIDD) and the CND, January 2022, https://cnduk.org/resources/ nuclear-weapons-are-banned-what-does-this-mean-for-britain/ (last accessed 16 July 2023).

16. 'TPNW signature and ratification status', ICAN, 2023, https://www. icanw.org/signature_and_ratification_status (last accessed 16 July 2023).

17. Author interview with Rebecca Johnson, Zoom, 30 March 2023.

18. Susi Snyder, 'Rejecting risk', PAX, ICAN, Profundo, January 2022, https://www.icanw.org/101_investors_say_no_to_nuclear_weapons (last accessed 16 July 2023).

19. The Fossil Fuel Non-Proliferation Treaty Initiative, https://fossilfuel-treaty.org/ (last accessed 16 July 2023).

20. Author interview with Rebecca Johnson, Zoom, 30 March 2023.

21. The World Counts: 'Number of planet earths needed for our consumption'. https://www.theworldcounts.com/economies/global/effects-of-consumerism (last accessed 16 October 2016).

22. E. Elhacham et al., 'Global human-made mass exceeds all living bio-

mass', *Nature* 588 (December 2020), pp. 442–4, https://www.nature.com/articles/s41586–020–3010–5 (last accessed 16 July 2023).

23. See Erina Brown et al., 'The potential for a plastic recycling facility to release microplastic pollution and possible filtration remediation effectiveness', *Journal of Hazardous Materials Advances* 10 (May 2023), pp. 2772–4166, ps://www.sciencedirect.com/science/article/pii/S2772416623000803 (last accessed 16 July 2023); and Eva Corlett, 'Microplastics found in freshly fallen Antarctic snow for first time', *The Guardian*, 8 June 2022, ps://www.theguardian.com/world/2022/jun/08/microplastics-found-in-freshly-fallen-antarctic-snow-for-first-time (last accessed 16 July 2023).

24. International Energy Agency (IEA), *The Role of Critical Minerals in Clean Energy Transitions*, IEA, Paris, 2021, https://www.iea.org/reports/the-role-of-critical-minerals-in-clean-energy-transitions (last accessed 16 July 2023).

25. Robert F. Kennedy, 'Remarks at the University of Kansas, March 18, 1968', John F. Kennedy Presidential Library and Museum, https://www.jfklibrary.org/learn/about-jfk/the-kennedy-family/robert-f-kennedy/robert-f-kennedy-speeches/remarks-at-the-university-of-kansas-march-18–1968 (last accessed 16 July 2023).

26. Juhasz, 'The quest to defuse Guyana's carbon bomb'.

27. Author interview with Melinda Janki, London, 23 February 2023.

28. Jason Hickel, 'Why growth can't be green', *Foreign Policy*, 12 September 2018, ps://foreignpolicy.com/2018/09/12/why-growth-cant-be-green/ (last accessed 16 July 2023).

29. Jason Hickel et al., 'Degrowth can work—here's how science can help', *Nature* 612 (12 December 2022), pp. 400–3, https://www.nature.com/articles/d41586-022-04412-x (last accessed 16 July 2023).

30. 'Beyond growth: Pathways towards sustainable prosperity in the EU', Beyond Growth, 15–17 May 2023, https://www.beyond-growth-2023.eu/ (last accessed 16 July 2023).

31. Letter: 'A post-growth Europe critical to survive and thrive, urge over 400 civil society groups and experts', Friends of the Earth Europe, 15 May 2023, https://friendsoftheearth.eu/publication/post-growth-europe-letter/ (last accessed 16 July 2023).

32. Aitor Hernández-Morales, 'How Vienna took the stigma out of social housing', *Politico*, 30 June 2022, https://www.politico.eu/article/vienna-social-housing-architecture-austria-stigma/ (last accessed 16 July 2023).

33. Kontrast.at, 'Mondragón: One of Spain's largest corporations belongs to its workers', Scoop.me, 8 March 2020, https://scoop.me/mondragon-one-of-the-largest-corporations-in-spain-belongs-to-its-workers/ (last accessed 16 July 2023).

34. 'Circular economy and industrial sustainability', Mondragon Unibertsitatea, 2023, https://www.mondragon.edu/en/research-transfer/engineering-technology/research-and-transfer-groups/-/mu-inv-mapping/grupo/economia-circular-y-sostenibilidad-industrial (last accessed 16 July 2023).

35. Lottie Limb, 'It's official: France bans short-haul domestic flights in favour of train travel', euronews.green, 23 May 2023, https://www.euronews.com/green/2022/12/02/is-france-banning-private-jets-everything-we-know-from-a-week-of-green-transport-proposals (last accessed 16 July 2023).

36. A Pacek, B. Radcliff and M. Brockway, 'Well-being and the democratic state: How the public sector promotes human happiness', *Social Indicators Research* 143 (2019), pp. 1147–59, https://link.springer.com/article/10.1007/s11205-018-2017-x (last accessed 16 July 2023).

37. 'Margaret Thatcher: A life in quotes', *The Guardian*, 8 April 2018, https://www.theguardian.com/politics/2013/apr/08/margaret-thatcher-quotes (last accessed 16 July 2023). The quote comes from an interview with Douglas Keay for *Woman's Own* on 23 September 1987.

38. Press release: 'Richest 1% bag nearly twice as much wealth as the rest of the world put together over the past two years', Oxfam International, 16 January 2023, https://www.oxfam.org/en/press-releases/richest-1-bag-nearly-twice-much-wealth-rest-world-put-together-over-past-two-years (last accessed 16 July 2023).

39. See 'World Happiness Report 2023: Happiest countries prove resilient despite overlapping crises', Wellbeing Research Centre, University of Oxford, 2023, https://wellbeing.hmc.ox.ac.uk/article/world-happiness-report-2023 (last accessed 16 July 2023); and Jonathan Este, 'Why Finland is the happiest country in the world—an expert explains', *The*

Conversation, 5 April 2023, https://theconversation.com/why-finland-is-the-happiest-country-in-the-world-an-expert-explains-203016 (last accessed 16 July 2023).

40. Eric Bergman, 'Facts and feelings: Do taxes make Finnish people happy?', This is Finland, March 2020, https://finland.fi/life-society/facts-and-feelings-do-taxes-make-finnish-people-happy/ (last accessed 16 July 2023).

41. Fintan Smith, 'Three quarters of Britons support wealth taxes on millionaires', YouGov, 23 January 2023, https://yougov.co.uk/topics/politics/articles-reports/2023/01/23/three-quarters-britons-support-wealth-taxes-millio (last accessed 16 July 2023).

42. '@fossilfreelondon crashed oil giant Shell's AGM this morning.' Tiktok video, Greenpeace UK, 23 May 2023, https://www.tiktok.com/@greenpeaceuk/video/7236382480427683099?_r=1&_t=8cpKJHSC3fW (last accessed 16 July 2023).

43. 'Shell hikes ex-CEO pay deal by 53% to £9.7m', Daily Business, 10 March 2023, ps://dailybusinessgroup.co.uk/2023/03/shell-hikes-ex-ceo-pay-deal-by-53-to-9-7m/ (last accessed 16 July 2023).

44. Jillian Ambrose, 'Shell accused of "profiteering bonanza" after record first-quarter profits of $9.6bn', *The Guardian*, 4 May 2023, https://www.theguardian.com/business/2023/may/04/shell-makes-record-quarterly-profits-of-nearly-10bn (last accessed 16 July 2023).

45. Shell plc, *Powering Progress: Shell Annual Report and Accounts for the Year 2022*, p. 16. Also available at https://reports.shell.com/annual-report/2022/ (last accessed 22 July 2023).

46. Andy Gregory, 'Shell company directors earned enough in 2021 to pay 390 nurses, as firm sees record profits', *The Independent*, 2 February 2023, https://www.independent.co.uk/news/business/shell-executives-pay-profits-who-b2274340.html (last accessed 22 July 2023).

47. '"You put safety second in pursuit of profits"—Caroline Dennett at Shell's AGM', NADJA, 13 June 2023, ps://www.nadja.co/2023/06/13/you-put-safety-second-in-pursuit-of-profits-caroline-dennett-at-shells-agm/ (last accessed 22 July 2023).

48. Sam Meredith, 'French riot police fire tear gas at climate protesters outside TotalEnergies shareholder meeting', CNBC, 26 May 2023,

https://www.cnbc.com/2023/05/26/totalenergies-agm-french-police-fire-tear-gas-at-climate-protesters.html (last accessed 22 July 2023).

49. Just Stop Oil (@JustStop_Oil), Twitter, 27 May 2023, 4:11 PM, https://twitter.com/JustStop_Oil/status/1662476684736573447?s=20 (last accessed 22 July 2023).

50. Peter Walker, 'Labour confirms plans to block all new North Sea oil and gas projects', *The Guardian*, 28 May 2023, https://www.theguardian.com/politics/2023/may/28/labour-confirms-plans-to-block-all-new-north-sea-oil-and-gas-projects (last accessed 22 July 2023).

51. Jeanne Theoharis, 'MLK would never shut down a freeway, and 6 other myths about the civil rights movement and Black Lives Matter', The Root, 15 July 2016, https://www.theroot.com/mlk-would-never-shut-down-a-freeway-and-6-other-myths-1790856033 (last accessed 22 July 2023).

52. James Ozden and Sam Glover, 'Literature review: Protest outcomes', Social Change Lab, April 2022, https://www.socialchangelab.org/_files/ugd/503ba4_21103ca09bf247748a788c92c60371a0.pdf (last accessed 22 July 2023).

53. Herbert H. Haines, 'Black radicalization and the funding of civil rights: 1957–1970', *Social Problems* 32, no. 1 (October 1984), pp. 31–43, https://academic.oup.com/socpro/issue/32/1 (last accessed 22 July 2023).

54. James Ozden and Sam Glover, 'Literature Review: Protest movement success factors', Social Change Lab, October 2022, p. 19, https://www.socialchangelab.org/_files/ugd/503ba4_e21c47302af942878411e-ab654fe7780.pdf (last accessed 22 July 2023).

55. James Ozden and Markus Ostarek, 'The radical flank effect of Just Stop Oil', Social Change Lab, December 2022, https://www.socialchangelab.org/_files/ugd/503ba4_a184ae5bbce24c228d07eda25566dc13.pdf (last accessed 22 July 2023).

56. Zoe Broughton, video recording of Lord Deben; Just Stop Oil (@JustStop_Oil), Twitter, 24 June 2023, 11:57 AM, https://twitter.com/JustStop_Oil/status/1672559621964279808 (last accessed 22 July 2023).

57. Fiona Harvey, 'UK missing climate targets on nearly every front, say government's advisers', *The Guardian*, 28 June 2023, https://www.the-

guardian.com/technology/2023/jun/28/uk-has-made-no-progress-on-climate-plan-say-governments-own-advisers (last accessed 22 July 2023).

58. Lancet Editorial, 'Doctors and Civil Disobedience', *The Lancet* 395 (25 January 2020), p. 248.

59. Letter to the author from Chris Newman, 7 June 2023.

60. Ozden and Glover, 'Literature review: Protest movement success factors', p. 18.

61. Damien Gayle, 'Climate diplomacy is hopeless, says author of How to Blow Up a Pipeline', *The Guardian*, 21 April 2023, https://www.theguardian.com/environment/2023/apr/21/climate-diplomacy-is-hopeless-says-author-of-how-to-blow-up-a-pipeline-andreas-malm (last accessed 22 July 2023).

62. Author interview with Indigo Rumbelow, London, 22 April 2023.

63. See Nina Lakhani, 'United Nations adopts landmark resolution on climate justice', *The Guardian*, 29 March 2023, https://www.theguardian.com/environment/2023/mar/29/united-nations-resolution-climate-emergency-vanuatu (last accessed 22 July 2023); and Press release: 'The General Assembly of the United Nations requests an advisory opinion from the Court on the obligations of States in respect of climate change', International Court of Justice, 19 April 2023, https://www.icj-cij.org/sites/default/files/case-related/187/187-20230419-PRE-01-00-EN.pdf (last accessed 22 July 2023).

64. Ruth Hayhurst, 'Court rules that campaigners can challenge new oil and gas licences', Drill or Drop?, 25 April 2023, https://drillordrop.com/2023/04/25/court-rules-that-campaigners-can-challenge-new-oil-and-gas-licences/ (last accessed 22 July 2023).

65. Central New Agency, 'Climate activist who allegedly held sign directed at jurors may be charged', *The Guardian*, 4 April 2023, https://www.theguardian.com/uk-news/2023/apr/04/climate-activist-trudi-warner-held-sign-telling-jurors-act-conscience-charged (last accessed 8 September 2023).

66. Press release: 17 May 2023: 'Judge Silas "Silencing" Reid defeated by show of solidarity for retired social worker arrested for holding up a sign': https://planb.earth/wp-content/uploads/2023/05/PR-Judge-Reid-defeated.pdf (last accessed 23 October 2023).

67. Press release: 'Judge Reid strikes back: Judge refers doctors, priest and Olympic Gold medalist to Attorney General over alleged contempt of court', Plan B, 2 June 2023, https://planb.earth/judge-reid-strikes-back/ (last accessed 4 August 2023).

68. 'Solidarity in contempt', Real Media, 18 May 2023, https://realmedia.press/solidarity-in-contempt/ (last accessed 22 July 2023).

69. Sofia Menchu, 'Guatemala court sentences journalist to 6 years in prison for money laundering', Reuters, 15 June 2023, https://www.reuters.com/world/americas/guatemala-court-sentences-journalist-6-years-money-laundering-2023–06–14/ (last accessed 7 September 2023).

70. Video of press conference, Maya UkuxBe, 19 April 2023, https://www.facebook.com/maria.ukux.5/videos/710003637539463 (last accessed 22 July 2023).

71. Press Release: 'Fig leaf falls from legal profession, as Bar Council Ethics Committee chair concedes barristers may act on conscience', Lawyers Are Responsible, 26 April 2023, https://www.lar.earth/press/press-release-26th-april-fig-leaf-falls-from-legal-profession-as-bar-council-ethics-committee-chair-concedes-barristers-may-act-on-conscience/ (last accessed 22 July 2023).

72. Sandra Laville, 'C of E divests of fossil fuels as oil and gas firms ditch climate pledges', *The Guardian* 22 June 2023, https://www.theguardian.com/world/2023/jun/22/c-of-e-divests-of-fossil-fuels-as-oil-and-gas-firms-ditch-climate-pledges (last accessed 26 November 2023).

73. Tessa Solomon, 'British Museum ends 27-year sponsorship deal with BP', ARTnews, 2 June 2023, https://www.artnews.com/art-news/news/british-museum-ends-bp-partnership-1234670391/ (last accessed 22 July 2023).

74. Press release: 'Mounting opposition to EACOP: New finance risk update highlights risks for banks and insurers', Insure Our Future, 17 April 2023, https://global.insure-our-future.com/mounting-opposition-to-eacop-new-finance-risk-update-highlights-risks-for-banks-and-insurers/ (last accessed 22 July 2023).

75. John Okot, '"Enemies of the state": Uganda targets climate activists in quiet crackdown', African Arguments, 16 March 2023, https://afri-

canarguments.org/2023/03/enemies-of-the-state-uganda-targets-activists-spearheading-fight-against-controversial-eacop-pipeline/ (last accessed 22 July 2023).

76. #STOPEACOP, https://www.stopeacop.net/home (last accessed 22 July 2023).

77. Amy McDonnell, 'Success in St Ives: Derek Thomas MP backs the climate & ecology bill', Zero Hour, 24 February 2023, https://www.zerohour.uk/success-in-st-ives/ (last accessed 22 July 2023).

78. 'Tackling the climate & nature crisis together', Zero Hour, 2023, https://www.zerohour.uk/bill/ (last accessed 22 July 2023).

79. They Work for You, 'How Derek Thomas voted on measures to prevent climate change', https://www.theyworkforyou.com/mp/25440/derek_thomas/st)ives/divisions?policy=1030 (last accessed 3 September 2023).

80. Lee Trewhela, 'Cornwall council votes to "reverse the destruction of nature" by backing the Climate & Ecology Bill', Cornwall Live, 19 April 2023, https://www.cornwalllive.com/news/cornwall-news/cornwall-council-votes-reverse-destruction-8364653 (last accessed 22 July 2023).

81. Rupert Read, 'Introducing the Climate Majority Project', YouTube, 16 March 2023, https://www.youtube.com/watch?v=sKKeLlRi9j0 (last accessed 22 July 2023).

82. 'How it began', Sustainable Penzance, 2023, https://sustainablepz.co.uk/about-us/ (last accessed 22 July 2023).

83. 'Climate emergency', Penzance Council, 2023, https://www.penzance-tc.gov.uk/climate-emergency/ (last accessed 22 July 2023).

84. Chenoweth, *Civil Resistance*, p. 251.

85. Letter to the author from Chris Newman, 7 June 2023.

86. Office for National Statistics, 'Worries about climate change, Great Britain: September to October 2022', 28 October 2022, https://www.ons.gov.uk/peoplepopulationandcommunity/wellbeing/articles/worriesaboutclimatechangegreatbritain/septembertooctober2022 (last accessed 22 July 2023).

87. Greta Thunberg, 'What we must do now', in Thunberg (ed.), *The Climate Book*, p. 327.

EPILOGUE

1. Energy Intelligence Forum 2023, 'Energy in a divided world', https://www.energyintelligenceforum.com/ (last accessed 25 October 2023).
2. Damien Gayle, 'Greta Thunberg arrested at London oil summit protest', *The Guardian*, 17 October 2023, https://www.theguardian.com/business/2023/oct/17/greta-thunberg-arrested-at-london-oil-summit-protest (last accessed 25 October 2023); and Kit Million Ross, 'Saudi Aramco boss claims renewables can't meet global power demand', Offshore Technology, 18 October 2023, https://www.offshore-technology.com/news/saudi-aramco-boss-claims-renewables-cant-meet-global-power-demand (last accessed 25 October 2023).
3. UN Environment Programme, 'The ocean is hotter than ever. Here's why', 22 August 2023, https://www.unep.org/news-and-stories/story/ocean-hotter-ever-heres-why (last accessed 25 October 2023).
4. National Snow and Ice Data System, 'Antarctic sets a record low maximum by wide margin', 25 September 2023, https://nsidc.org/arctic-seaicenews/2023/09/antarctic-sets-a-record-low-maximum-by-wide-margin/ (last accessed 25 October 2023).
5. NOAA National Centers for Environmental Information, 'Global climate summary for September 2023', Climate.gov, 16 October 2023, https://www.climate.gov/news-features/understanding-climate/global-climate-summary-september-2023 (last accessed 25 October 2023); and Damian Carrington, '"Gobsmackingly bananas": scientists stunned by planet's record September heat', *The Guardian*, 5 October 2023, https://www.theguardian.com/environment/2023/oct/05/gobsmackingly-bananas-scientists-stunned-by-planets-record-september-heat (last accessed 25 October 2023).
6. UN News, 'Hottest July ever signals "era of global boiling has arrived" says UN chief', 27 July 2023, https://news.un.org/en/story/2023/07/1139162 (last accessed 25 October 2023).
7. Sam Ezra Fraser-Baxter, 'Scientists uncover the role of climate change in devastating East Canada fires', Imperial College London, 24 August 2023, https://www.imperial.ac.uk/news/247154/scientists-uncover-role-climate-change-devastating/ (last accessed 25 October 2023); and Evan Bush, 'Florida hit by Canadian smoke from historic wildfire season',

NBC News, 3 October 2023, https://www.nbcnews.com/science/environment/florida-canada-wildfire-smoke-haze-rcna118701 (last accessed 25 October 2023).

8. World Weather Attribution, 'Interplay of climate change-exacerbated rainfall, exposure and vulnerability led to widespread impacts in the Mediterranean region', 19 September 2023, https://www.worldweatherattribution.org/interplay-of-climate-change-exacerbated-rainfall-exposure-and-vulnerability-led-to-widespread-impacts-in-the-mediterranean-region/ (last accessed 25 October 2023).

9. Jonathan Watts et al., 'The hottest summer in human history-A visual timeline', *The Guardian*, 29 September, 2023, https://www.theguardian.com/environment/ng-interactive/2023/sep/29/the-hottest-summer-in-human-history-a-visual-timeline?CMP=share_btn_tw (last accessed 25 October 2023).

10. Press release: 'Tourism, Italy: no climate emergency, temperatures in seasonal norm', ENIT Agenzia Nazionale del Turismo, 21 July 2023, https://www.enit.it/en/tourism-italy-no-climate-emergency-temperatures-in-seasonal-norm (last accessed 25 October 2023).

11. Katherine Richardson et al., 'Earth beyond six of nine planetary boundaries' *ScienceAdvances* 9, no. 37 (2023), https://www.science.org/doi/10.1126/sciadv.adh2458 (last accessed 25 October 2023).

12. Dunja Mijatović, 'Crackdowns on peaceful environmental protests should stop and give way to more social dialogue', Council of Europe, 2 June 2023, https://www.coe.int/en/web/commissioner/-/crackdowns-on-peaceful-environmental-protests-should-stop-and-give-way-to-more-social-dialogue (last accessed 25 October 2023).

13. Damien Gayle, 'Just Stop Oil protesters have appeals blocked over Dartford crossing sentences', *The Guardian*, 31 July 2023, https://www.theguardian.com/environment/2023/jul/31/just-stop-oil-protesters-have-appeals-blocked-over-dartford-crossing-sentences (last accessed 25 October 2023); and Shivani Chaudhari and Tom Pilgrim, 'Just Stop Oil protesters bid to challenge jail terms refused', BBC News, 12 October 2023, https://www.bbc.co.uk/news/uk-england-essex-6707 9796 (last accessed 25 October 2023).

14. Anita Mureithi, 'Climate activists face "crippling" legal fees for injunc-

tions banning protest', Open Democracy, 20 September 2023, https://www.opendemocracy.net/en/civil-injunctions-just-stop-oil-insulate-britain-extinction-rebellion-punishment-protests/ (last accessed 25 October 2023).

15. Press release: 'UK public bodies banned from imposing their own boycotts against foreign countries', GOV.UK, 19 June 2023, https://www.gov.uk/government/news/uk-public-bodies-banned-from-imposing-their-own-boycotts-against-foreign-countries (last accessed 25 October 2023).

16. Press release: 'Protect the right to boycott unethical companies, Quakers urge MPs', Quakers in Britain, 21 June 2023, https://www.quaker.org.uk/news-and-events/news/protect-the-right-to-boycott-unethical-companies-quakers-urge-mps (last accessed 25 October 2023).

17. Josiah Mortimer, 'Activist Who Held Sign Outside Court Informing Jurors of "Right to Acquit" Insulate Britain Activists Could Face Two Years in Jail', Byline Times, 18 September 2023, https://bylinetimes.com/2023/09/18/activist-who-held-sign-outside-court-informing-jurors-of-right-to-acquit-insulate-britain-activists-could-face-two-years-in-jail/ (last accessed 25 October 2023).

18. Patrick Daly, 'Sunak says blanket 20mph zones and low traffic neighbourhoods "need to stop"', *Independent*, 30 September 2023, https://www.independent.co.uk/news/uk/politics/rishi-sunak-20mph-zones-cars-b2421291.html (last accessed 25 October 2023).

19. BBC News, 'Rosebank oil field: What is the row over the project?', 27 September 2023, https://www.bbc.co.uk/news/business-66933832 (last accessed 25 October 2023).

20. UNFCCC. Secretariat, 'Technical dialogue of the first global stocktake. Synthesis report by the co-facilitators on the technical dialogue', UN Framework Convention on Climate Change, 8 September 2023, https://unfccc.int/documents/631600 (last accessed 25 October 2023).

21. Carbon Brief, 'In-depth Q&A: What do Rishi Sunak's U-turns mean for UK climate policy?', 22 September 2023, https://www.carbonbrief.org/in-depth-qa-what-do-rishi-sunaks-u-turns-mean-for-uk-climate-policy/ (last accessed 25 October 2023).

22. UN News, '"Humanity has opened the gates to hell" warns Guterres

as climate coalition demands action', 20 September 2023, https://news. un.org/en/story/2023/09/1141082 (last accessed 25 October 2023).

23. Matthew Taylor and Pamela Duncan, 'Revealed: almost everyone in Europe is breathing toxic air', *The Guardian*, 20 September 2023, https://www.theguardian.com/environment/2023/sep/20/revealed-almost-everyone-in-europe-breathing-toxic-air (last accessed 25 October 2023).

24. These short quotes are taken from Rishi Sunak's speech to the Conservative Party conference in Manchester, October 2023. Full speech available here: Conservative Home, '"Be in no doubt it is time for a change and we are it." Sunak's conference speech – full text', 4 October 2023, https://conservativehome.com/2023/10/04/be-in-no-doubt-it-is-time-for-a-change-and-we-are-it-sunaks-conference-speech-full-text/ (last accessed 25 October 2023).

25. Press release: 'New survey reveals British public generally think disruptive, non-violent protesters should not be imprisoned,' School of Psychological Science, University of Bristol, 1 August 2023, https://www.bristol.ac.uk/psychology/news/2023/116.html (last accessed 25 October 2023).

26. Helena Horton, 'Only 22% of Britons trust Sunak on climate', *The Guardian*, 23 September 2023, https://www.theguardian.com/environment/2023/sep/23/only-22-percent-of-britons-trust-sunak-on-climate-finds-guardian-poll (last accessed 25 October 2023).

27. Welsh Government, 'Safer at 20mph: Let's look out for each other', https://www.gov.wales/safer-20mph-lets-look-out-each-other (last accessed 25 October 2023); and Xander Elliards, 'Humza Yousaf joins condemnation of UK Government's Rosebank approval', *The National*, 27 September 2023, https://www.thenational.scot/news/23817127.humza-yousaf-joins-condemnation-uk-governments-rosebank-approval/ (last accessed 25 October 2023).

28. The Tennis Letter (@TheTennisLetter), Twitter, 8 September 2023, 4:51 AM, https://twitter.com/thetennisletter/status/1699993832119 832809?s=48 (last accessed 25 October 2023).

29. Maddy Shaw Roberts, 'Conductor Vladimir Jurowski tells audience to "let climate protestors speak" as they interrupt Bruckner symphony',

Classic FM, 11 September 2023, https://www.classicfm.com/composers/bruckner/vladimir-jurowski-audience-climate-protestors-speak/ (last accessed 25 October 2023).

30. Bianca Castro, 'Protesters vow to repeat juror campaign outside court', The Law Society Gazette, 25 September 2023, https://www.lawgazette.co.uk/news/protesters-vow-to-repeat-juror-campaign-outside-court/5117338.article (last accessed 25 October 2023); and Richard Vogler, 'Trudi Warner reveals the dark secret of English courts: juries do have the right to follow their consciences', The Guardian, 27 September 2023, https://www.theguardian.com/commentisfree/2023/sep/27/trudi-warner-english-courts-juries (last accessed 10 November 2023).

31. Andrew Lee, 'As another offshore wind chief quits Shell, was there a social media parting shot at CEO?', Recharge, 7 July 2023, https://www.rechargenews.com/wind/as-another-offshore-wind-chief-quits-shell-was-there-a-social-media-parting-shot-at-ceo-/2-1-1483062 (last accessed 25 October 2023).

32. Damien Gayle, 'Joe Lycett backs out of award ceremony over fossil fuel links', The Guardian, 20 June 2023, https://www.theguardian.com/culture/2023/jun/20/exclusive-joe-lycett-backs-out-award-ceremony-over-fossil-fuel-links (last accessed 26 October 2023).

33. Daniel Binns, 'Author Mikaela Loach walks out of her Edinburgh book festival event over fossil fuel links', Sky News, 13 August 2023, https://news.sky.com/story/author-mikaela-loach-walks-out-of-her-edinburgh-book-festival-event-over-fossil-fuel-links-12939634 (last accessed 26 October 2023).

34. Sara Orchard, 'Rugby World Cup 2023: Oil companies should not sponsor World Cups, says David Pocock', BBC Sport, 7 September 2023, https://www.bbc.co.uk/sport/rugby-union/66723726 (last accessed 26 October 2023).

35. Press release: 'Which? puts major high-street banks in "red" warning category based on green credentials', Which?, 5 October 2023, https://www.which.co.uk/policy-and-insight/article/which-puts-major-high-street-banks-in-red-warning-category-based-on-green-credentials-apzxi2Y8oFqz (last accessed 26 October 2023).

36. Katie Surma, 'After Decades of Oil Drilling, Indigenous Waorani Group

Fights New Industry Expansions In Ecuador', Inside Climate News, 30 August 2023, https://insideclimatenews.org/news/30082023/ecuador-votes-oil-ban-on-indigenous-land/ (last accessed 26 October 2023).

37. Vanessa Buschschlüter, 'Guatemala paralysed as pro-democracy protests run into second week', BBC News, 10 October 2023, https://www.bbc.co.uk/news/world-latin-america-67064814 (last accessed 26 October 2023); and see https://www.facebook.com/photo/?fbid=648766320717896&set=pcb.648766610717867 (last accessed 26 October 2023).

38. *Extinction Rebellion Netherlands*, 'A12 blockades Netherlands effective: Lower House asks cabinet for phase-out plan for fossil fuel subsidies', *10 October 2023*, https://extinctionrebellion.nl/en/a12-blockades-netherlands-effective-lower-house-asks-cabinet-for-phase-out-plan-for-fossil-fuel-subsidies/ (last accessed 26 October 2023).

39. James Gregory, 'Chris Packham to challenge Rishi Sunak over net zero policy delays', BBC News, 4 October 2023, https://www.bbc.co.uk/news/uk-67008481 (last accessed 26 October 2023).

40. Liberty, 'Court grants permission for Liberty legal action against Home Secretary', 4 October 2023, https://www.libertyhumanrights.org.uk/issue/court-grants-permission-for-liberty-legal-action-against-home-secretary/ (last accessed 26 October 2023).

41. Lesley Clark, 'Kids Sued Montana over Climate Change and Won', *Scientific American*, 15 August 2023, https://www.scientificamerican.com/article/kids-sued-montana-over-climate-change-and-won/ (last accessed 26 October 2023).

42. Global Legal Action Network, 'The Youth4ClimateJustice hearing is over, but the work continues!', Crowd Justice, 12 October 2023, https://www.crowdjustice.com/case/youth4climatejustice/ (last accessed 26 October 2023).

43. Adrian Murdoch, 'Funding fossil fuels still pays', Capital Monitor, 20 April 2023, https://capitalmonitor.ai/asset-class/fixed-income/funding-fossil-fuels-still-pays/ (last accessed 26 October 2023).

44. Patrick Greenfield and Jonathan Watts, 'JP Morgan economists warn climate crisis is threat to human race', *The Guardian*, 21 February 2020, https://www.theguardian.com/environment/2020/feb/21/jp-morgan-

economists-warn-climate-crisis-threat-human-race (last accessed 26 October 2023).

45. Richard Horton, 'Offline: We have a decade', *The Lancet* 402, no. 10408 (2023), https://www.thelancet.com/journals/lancet/article/PIIS0140-6736(23)02181-5/fulltext (last accessed 26 October 2023).

46. Chris Smyth, Eleanor Hayward and Oliver Wright, 'Covid inquiry: Matt Hancock admits pandemic plan was totally wrong—as it happened', *The Times*, 27 June 2023, https://www.thetimes.co.uk/article/covid-inquiry-matt-hancock-latest-news-live-2023-kt3hvbj6x (last accessed 10 November 2023).

SELECT BIBLIOGRAPHY

Adams, R. [Lynne Jones]. 'Dialogue across the Iron Curtain', *New Statesman*, 12 October 1984, pp. 21–2.

———. 'Take not that hypocritic oath', *New Statesman*, 4 April 1986, p. 19.

Ambrose, J. 'Shell accused of "profiteering bonanza" after record first-quarter profits of $9.6bn', *The Guardian*, 4 May 2023, https://www.theguardian.com/business/2023/may/04/shell–makes-record-quarterly-profits-of-nearly-10bn

Ascherson, N. *The Struggles for Poland*. New York: Random House, 1987.

Atkinson, J. [Lynne Jones]. 'The woman behind Solidarity: The story of Anna Walentynowicz', *Ms Magazine* 12, no. 8 (February 1984), pp. 96–8.

Ballester J., et al. 'Heat-related mortality in Europe during the summer of 2022', *Nature Medicine* 29 (July 2023), pp. 1857–66, https://doi.org/10.1038/s41591-023-02419-z

Bamford, S. *Passages in the Life of a Radical, and Early Days*, vol. 2. London: T. Fisher Unwin, 1841.

Barnett, D. 'The absurdity of owning moors and mountains', *The Independent*, 27 April 2022, https://www.independent.co.uk/independentpremium/long-reads/owner-moors-mountains-kinder-scout-mass-trespass-b2063359.html

Bearman, C. J. 'An examination of Suffragette violence', *English Historical Review* 120, no. 486 (April 2005), pp. 365–97, https://doi.org/10.1093/ehr/cei119

Blaisdell, B. (ed.), *Essays on Civil Disobedience*. New York: Dover Publications, 2016.

Bregman, R. *Humankind: A Hopeful History*. London: Bloomsbury, 2020.

Brown, E., et al. 'The potential for a plastic recycling facility to release microplastic pollution and possible filtration remediation effectiveness', *Journal of Hazardous Materials Advances* 10 (May 2023), pp. 2772–4166, https://doi.org/10.1016/j.hazadv.2023.100309

Butler, S. 'RNLI had a record year after Farage attacks and people are thrilled', *indy100*, 2 January 2022, https://www.indy100.com/news/rnli-record-year-donations-nigel-farage-b1985559

Campbell, B. *The Iron Ladies: Why do Women Vote Tory?* London: Virago, 1987.

Campbell, L. 'Revealed: Britain's worst banks for emissions (and the best)', Good with Money, 12 October 2022, https://good-with-money.com/2022/10/12/revealed-britains-worst-banks-for-emissions-and-the-best/

Carrington, D. 'Desmond Tutu calls for anti-apartheid style boycott of fossil fuel industry', *The Guardian*, 10 April 2014, https://www.the-guardian.com/environment/2014/apr/10/desmond-tutu-anti-apartheid-style-boycott-fossil-fuel-industry

Cassel, C. K. 'The Nevada desert demonstration', *Medicine and War* 3, no. 3 (1987), pp. 141–3, https://www.jstor.org/stable/45353120

Chenoweth, E. *Civil Resistance: What Everyone Needs to Know*. New York: Oxford University Press, 2021.

———. 'People power', in G. Thunberg (ed.), *The Climate Book*. London: Allen Lane, 2022, pp. 364–8.

Cuff, M. 'How can we keep homes cool in extreme heat without air conditioning?', *New Scientist*, 13 July 2023, https://www.newscientist.com/article/2382452-how-can-we-keep-homes-cool-in-extreme-heat-without-air-conditioning/

Davies, D. *Heart of Europe: A Short History of Poland*. Oxford: Oxford University Press, 1986.

Dickens, C. 'The Letters: 1856–1858 (1858): To Miss Burdett-Coutts, 4 October 1857', in G. Storey and K. M. Tillotson (eds.), *The British Academy/The Pilgrim Edition of the Letters of Charles Dickens, Vol. 8:*

1856–1858. Oxford: Oxford University Press, 1995, https://dx.doi.org/10.1093/oseo/instance.00161597

Douzinas, C., S. Homewood and R. Warrington. 'The shrinking scope for public protest', *Index on Censorship* 17, no. 8 (1988), pp. 12–15, https://doi.org/10.1080/03064228808534499

Driver, C. *The Disarmers: A Study in Protest*. London: Hodder and Stoughton, 1964.

Dwyer R. 'State archives: Dunnes Stores strike demonstrated power of the few', *The Irish Examiner*, 1 January 2016, https://www.irishexaminer.com/news/arid-20373917.html

Elhacham, E., et al. 'Global human-made mass exceeds all living biomass', *Nature* 588 (December 2020), pp. 442–4, https://doi.org/10.1038/s41586-020-3010-5

END. 'The Appeal for European Nuclear Disarmament', Bertrand Russell Peace Foundation, April 1980, http://www.russfound.org/END/EuropeanNuclearDisarmament.html

Extinction Rebellion. *This is Not a Drill: An Extinction Rebellion Handbook*. London: Penguin, 2019.

Freedman, L. 'Thatcherism and defence', in D. Kavanagh and A. Seldon (eds.), *The Thatcher Effect*. Oxford: Clarendon Press, 1989, pp. 143–53.

Freeman, J. 'The tyranny of structurelessness' [1970], Jo Freeman.com, https://www.jofreeman.com/joreen/tyranny.htm

Gandhi, M. 'Satyagraha (Noncooperation) (1920)', in Blaisdell (ed.), *Essays on Civil Disobedience*, pp. 92–4.

Gayle, D. 'Climate diplomacy is hopeless, says author of How to Blow Up a Pipeline', *The Guardian*, 21 April 2023, https://www.theguardian.com/environment/2023/apr/21/climate-diplomacy-is-hopeless-says-author-of-how-to-blow-up-a-pipeline-andreas-malm

General Medical Council. *Good Medical Practice*, GMC, 29 April 2019, https://www.gmc-uk.org/ethical-guidance/ethical-guidance-for-doctors/good-medical-practice/duties-of-a-doctor

Gregoire, P. 'Australia has a long history of fighting harm minimization drug policies', *Vice*, 11 January 2016, https://www.vice.com/en/article/4wbevq/how-australia-led-the-world-in-progressive-drug-policy-then-went-backwards

SELECT BIBLIOGRAPHY

Guatemalan Commission for Historical Clarification (CEH). *Guatemala, Memory of Silence: Report of the Commission for Historical Clarification* (1999), pp. 17–23.

Haines, H. H. 'Black radicalization and the funding of civil rights, 1957–1970', *Social Problems* 32, no. 1 (October 1984), pp. 31–43.

Harvey, F. 'UK missing climate targets on nearly every front, say government's advisers', *The Guardian*, 28 June 2023, https://www.theguardian.com/technology/2023/jun/28/uk-has-made-no-progress-on-climate-plan-say-governments-own-advisers

Heyrick, E. *Immediate, Not Gradual Abolition, or, An Inquiry into the Shortest, Safest, and Most Effectual Means of Getting Rid of West Indian Slavery*. Boston: Isaac Knapp, 1838. Available at Library of Congress: African American Pamphlet Collection copy, pp. 4, 8, 34, https://www.loc.gov/item/11009325/

Hickel, J. 'Why growth can't be green', *Foreign Policy*, 12 September 2018, https://foreignpolicy.com/2018/09/12/why-growth-cant-be-green/

Hickel, J., at al. 'Degrowth can work—here's how science can help', *Nature* 612 (12 December 2022), pp. 400–3, https://www.nature.com/articles/d41586-022-04412-x

Hickman, C., et al. 'Climate anxiety in children and young people and their beliefs about government responses to climate change: A global survey', *The Lancet Planetary Health* 5, no. 12 (2021), e863–73, https://doi.org/10.1016/S2542-5196(21)00278-3

Hipperson, S. *Greenham: Non-Violent Women v The Crown Prerogative*. London: Greenham Publications, 2005.

Howden, D. 'Europe's new anti-migrant strategy? Blame the rescuers', *Prospect* (20 March 2018), https://www.prospectmagazine.co.uk/magazine/europes-new-anti-migrant-strategy-blame-the-rescuers

Hyde, J. 'Lawyers should have to warn clients about environmental damage, say campaigners', *Law Society Gazette*, 16 September 2022, https://www.lawgazette.co.uk/news/lawyers-should-have-to-warn-clients-about-environmental-damage-say-campaigners/5113693.article

International Energy Agency (IEA). *The Role of Critical Minerals in Clean Energy Transitions*. Paris, May 2021, https://www.iea.org/reports/the-role-of-critical-minerals-in-clean-energy-transitions

SELECT BIBLIOGRAPHY

Johnson, R. 'The Nuclear Ban and Humanitarian Strategies to Eliminate Nuclear Threats' in B. N. V. Steen and O. Njølstad (eds.), *Nuclear Disarmament: A Critical Assessment*. Abingdon: Routledge, 2019, pp. 75–93.

Jones, L. 'Breaking barriers', *New Statesman*, October 1983, p. 14.

———. 'In the eye of the storm', *New Statesman*, 16 December 1983, pp. 8–9.

———. 'On common ground: The women's peace camp at Greenham Common', in L. Jones (ed.), *Keeping the Peace*. London: Women's Press, 1983, pp. 79–97.

———. 'Shut up and listen', *New Statesman*, 2 March 1984, pp. 10–11.

———. 'Changing ideas of authority', *New Statesman*, 20 November 1984, pp. 8–9.

———. 'Letter to a woman on the Falklands Victory Parade, October 1982', in B. Harford and S. Hopkins (eds.), *Greenham Common: Women at the Wire*. London: Women's Press, 1984, pp. 74–7.

———. *Outside the Asylum: A Memoir of War, Disaster and Humanitarian Psychiatry*. London: Weidenfeld and Nicolson, 2017.

———. *The Migrant Diaries*. New York: Refuge Press, 2021.

Joseph Rowntree Foundation. 'UK poverty 2023: The essential guide to understanding poverty in the UK'. Joseph Rowntree Foundation, January 2023, https://www.jrf.org.uk/report/uk-poverty-2023#key-findings

Juhasz, A. 'The quest to defuse Guyana's carbon bomb', Wired, 20 December 2022, https://www.wired.com/story/the-quest-to-defuse-carbon-bomb-guyana/

Kaminski, J. P., et al. (eds.), *The Documentary History of the Ratification of the Constitution Digital Edition*. Charlottesville: University of Virginia Press, 2009, https://rotunda.upress.virginia.edu/founders/RNCN.html

King, M. L. 'Letter from Birmingham Jail (1963)', in Blaisdell (ed.), *Essays on Civil Disobedience*, pp. 132–48.

———. 'Love, law, and civil disobedience (1961)', in Blaisdell (ed.), *Essays on Civil Disobedience*, pp. 120–31.

Lambert, D. 'A defence of criminal damage', RSA, 13 May 2021, https://www.thersa.org/comment/2021/05/a-defence-of-criminal-damage

Lancet Editorial. 'Doctors and civil disobedience', *The Lancet* 395

(25 January 2020), p. 248, https://doi.org/10.1016/S0140–6736(20) 30120–3

Lancet Planetary Health Editorial, 'A role for provocative protest', *Lancet Planetary Health* 6, no. 11 (November 2022), e846, https://doi.org/10.1016/S2542-5196(22)00287-X

Landphair, J. '"The forgotten people of New Orleans": Community, vulnerability, and the Lower Ninth Ward', *Journal of American History* 94, no. 3 (December 2007), pp. 837–45, https://doi.org/10.2307/25095146

Lawson, A. 'Consultant who ditched Shell: "Take a look at yourselves in the mirror"', *The Guardian*, 28 May 2022, https://www.theguardian.com/business/2022/may/28/consultant-who-ditched-shell-take-a-look-at-yourselves-in-the-mirror

Levy, M. *Ban the Bomb! Michael Randle and Direct Action against Nuclear War*. Stuttgart: Ibidem Press, 2021.

Liddington, J. *The Long Road to Greenham: Feminism and Anti-Militarism in Britain Since 1820*. London: Virago, 1989.

Marcou, A. 'Violence, communication, and civil disobedience', *Jurisprudence* 12, no. 4 (2021), pp. 491–511, https://doi.org/10.1080/20403313.2021.1921494

Mastnak, L. [Lynne Jones]. 'The process of engagement in non-violent collective action: Case studies from the 1980s', unpublished PhD thesis, Bath University, 1995.

Miller, T., N. Buxton and M. Akkerman. 'Global climate wall: How the world's wealthiest nations prioritise borders over climate action', Report by Transnational Institute, 25 October 2021, https://www.tni.org/en/publication/global-climate-wall

Needham, A. *The Hammer Blow: How Ten Women Disarmed a War Plane*. London: Peace News Press, 2016.

OECD, 'Innovative citizen participation and new democratic institutions: Catching the deliberative wave', Paris: OECD Publishing, June 2020, https://www.oecd.org/gov/innovative-citizen-participation-and-new-democratic-institutions-339306da-en.htm

Olusoga, D. *Black and British: A Forgotten History*. London: Macmillan, 2016.

Openshaw, S., P. Steadman and O. Greene. *Doomsday: Britain after Nuclear Attack*. Oxford: Blackwell, 1983.

SELECT BIBLIOGRAPHY

Oxfam International. 'Inequality kills', Oxfam Briefing Paper, Oxfam International, 17 January 2022, https://www.oxfam.org/en/research/inequality-kills

———. 'Survival of the richest', Oxfam Briefing Paper, Oxfam International, 16 January 2023, https://www.oxfam.org/en/research/survival-richest

Ozden J., and S. Glover. 'Literature review: Protest outcomes', Social Change Lab, April 2022, https://www.socialchangelab.org/_files/ugd/5 03ba4_21103ca09bf247748a788c92c60371a0.pdf

———. 'Literature Review: Protest movement success factors', Social Change Lab, October 2022, https://www.socialchangelab.org/_files/ugd/503ba4_e21c47302af942878411eab654fe7780.pdf

Ozden J., and M. Ostarek. 'The radical flank effect of Just Stop Oil', Social Change Lab, December 2022, https://www.socialchangelab.org/_files/ugd/503ba4_a184ae5bbce24c228d07eda25566dc13.pdf

Pacek, A., B. Radcliff and M. Brockway. 'Well-being and the democratic state: How the public sector promotes human happiness', *Social Indicators Research* 143 (2019), pp. 1147–59, https://doi.org/10.1007/s11205-018-2017-x

Pankhurst, E. '"Freedom or death": Speech delivered in Hartford, Connecticut on November 13, 1913', *The Guardian*, 27 April 2007, https://www.theguardian.com/theguardian/2007/apr/27/greatspeeches

Quinault, R. 'Gladstone and slavery', *Historical Journal* 52, no. 2 (June 2009), pp. 363–83.

Rai, M. 'Ziegler: Celebrating the Supreme Court decision one year on', *Peace News*, 1 June 2022, https://peacenews.info/node/10270/ziegler-celebrating-supreme-court-decision-one-year

———. 'Ziegler: The full story behind the ground-breaking Supreme Court decision', *Peace News*, 20 June 2022, https://peacenews.info/blog/2022/ziegler-full-story-behind-ground-breaking-supreme-court-decision

Randle, M. *Civil Resistance*. London: Fontana, 1994.

———. *Rebel Verdict*. Sparsnäs: Irene, 2022.

Rawls, J. 'The justification of civil disobedience', in S. Freeman (ed.), *John Rawls: Collected Papers*. Cambridge, MA: Harvard University Press, 1999, pp. 176–89.

SELECT BIBLIOGRAPHY

Raymond, J. 'Still carrying Greenham home', *The Land* 29 (2021), https://www.thelandmagazine.org.uk/articles/still-carrying-greenham-home

Reid, R. *The Peterloo Massacre*. Portsmouth, NH: Heinemann, 1989.

———. 'What next on climate? The need for a new moderate flank', *Perspectiva*, 6 October 2021, https://systems-souls-society.com/what-next-on-climate-the-need-for-a-moderate-flank/

Reznicek, J., and R. Montoya. 'Why we acted', *Via Pacis: The Voice of the Des Moines Catholic Worker Community* 41, no. 3 (October 2017), pp. 1, 3.

Riddell, F. *Death in Ten Minutes: The Forgotten Life of Radical Suffragette Kitty Marion*. London: Hodder and Stoughton, 2018, pp. 117–21.

———. 'Sanitising the suffragettes', *History Today* 68, no. 2 (February 2018), https://www.historytoday.com/history-matters/sanitising-suffragettes

RNLI. 'Statement on the humanitarian work of the RNLI in the English Channel', RNLI Lifeboats, 28 July 2021, https://rnli.org/news-and-media/2021/july/28/statement-on-the-humanitarian-work-of-the-rnli-in-the-english-channel

Rosenberg, M. B. *Nonviolent Communication: A Language of Compassion*. Del Mar, CA: Puddle Dancer Press, 1999.

Runciman, D. *How Democracy Ends*. London: Profile Books, 2018.

Rylance, M. 'Why I'm resigning from the RSC', Culture Unstained, 21 June 2019, https://cultureunstained.org/2019/06/21/mark-rylance-rsc/

Sanson, A., and M. Bellemo. 'Children and youth in the climate crisis', *BJPsych Bull*etin 45, no. 4 (August 2021), pp. 205–9, http://doi:10.1192/bjb.2021.16

Sharp, G. *The Politics of Nonviolent Action, Part One: Power and Struggle*. Boston: Porter Sargent, 1973.

———. *The Politics of Nonviolent Action, Part Two: The Methods of Nonviolent Action*. Boston: Porter Sargent, 1973.

———. *The Politics of Nonviolent Action, Part Three: The Dynamics of Nonviolent Action*. Boston: Porter Sargent, 1973.

Smoke, B. 'The Stansted 15's quashed conviction shows we were never terrorists', *The Guardian*, 2 February 2021, https://www.theguardian.com/commentisfree/2021/feb/02/stansted-15-quashed-conviction-terrorists-deportation-hostile-environment

Taylor, M. 'Royal Shakespeare Company to end BP sponsorship deal', *The Guardian*, 2 October 2019, https://www.theguardian.com/stage/2019/oct/02/royal-shakespeare-company-to-end-bp-sponsorship-deal

Taylor, R. *Against the Bomb: The British Peace Movement, 1958–1965.* Oxford: Oxford University Press, 1995.

Thompson, E. P. *The Making of the English Working Class.* London: Penguin, 1980.

———. *Protest and Survive.* Nottingham: Russell Press for the Campaign for Nuclear Disarmament and the Bertrand Russell Peace Foundation, 1980.

Thoreau, H. D. 'Civil disobedience (1849)', in Blaisdell (ed.), *Essays on Civil Disobedience*, pp. 22–42.

———. 'Slavery in Massachusetts (1854)', in Blaisdell (ed.), *Essays on Civil Disobedience*, pp. 43–56.

Thunberg, G. (ed.). *The Climate Book.* London: Allen Lane, 2022.

Tulchinsky, T. H. 'John Snow, cholera, the Broad Street Pump: Waterborne diseases then and now', *Case Studies in Public Health*, 30 March 2018, pp. 77–99, https://www.ncbi.nlm.nih.gov/pmc/articles/PMC7150208/

Vogler R. 'Trudi Warner reveals the dark secret of English courts: juries do have the right to follow their consciences', *The Guardian*, 27 September 2023, https://www.theguardian.com/commentisfree/2023sep/27/trudi-warner-english-courts-juries

Waal, F. de. *The Bonobo and the Atheist: In Search of Humanism among the Primates.* New York: W. W. Norton, 2013, pp. 228–40.

Washington, G. *George Washington's Farewell Address.* Carlisle: Applewood Books, 1999.

Watts, N., et al. 'The 2018 report of the Lancet Countdown on health and climate change: Shaping the health of nations for centuries to come', *The Lancet* 392 (2018), pp. 2479–514, https://doi.org/10.1016/S0140–6736(18)32594-7

Webb, S. *The Suffragette Bombers: Britain's Forgotten Terrorists.* Barnsley: Pen and Sword, 2014.

Wellbeing Research Centre. 'World Happiness Report 2023: Happiest countries prove resilient despite overlapping crises', Wellbeing Research

SELECT BIBLIOGRAPHY

Centre, University of Oxford, 2023, https://wellbeing.hmc.ox.ac.uk/article/world-happiness-report-2023

Wilde, O. 'The Soul of Man under Socialism' in L. Dowling (ed.), *The Soul of Man Under Socialism and Selected Critical Prose*. London: Penguin, 2001, pp. 125–162.

INDEX

INDEX

Airport expansion permission (2022), 199–200

Colston statue toppling (2020), 99–104, 155

St Pauls *riot* (1980), 101, 143

British Aerospace, 86–7, 92

British Broadcasting Corporation (BBC), 60, 68, 96, 115, 150, 243

British Film Institute, 239

British LGBT Awards, 324

British Medical Journal, 17, 114, 115, 149

British Museum, London, 236–7, 314

Brittan, Leon, 16

Brixton riots (1981), 1

bronchiectasis, 148

Brooks, Jeff, 294

Brown v. Board of Education (1954), 226

Bruckner, Anton, *323–4*

Bucher, Gabriela, 303

'building back better', 193

Bulletin of the Atomic Scientists, 296

Bunten, Anna, 148

BuzzFeed, 57

Caal Xol, Bernardo, 259, 260–72, 313, 325

cab rank rule, 245

caesarean sections, 122

Cahabón river, 260–72

Calais, France, 194

Camborne, Cornwall, 164

Cambridgeshire County Council, 93

Cameron, James, 109

Campaign for Nuclear Disarmament (CND), 12, 17, 21, 24, 58, 78, 248

Campbell, Dirk, 304

'Can Property Damage be Nonviolent?', 92

Canada, 320

Canary Wharf, London, 72–3, 152

Canning Town, London, 55–60, 69, 121, 157

carbon dioxide, 33, 63

Cardiff, Wales, 2, 18, 19, 293

Carlin, David, 235

Carney, Mark, 73, 96, 236

Carrera, Rafael, 280

Carter, James 'Jimmy', 9

Castañeda de León, Oliverio, 279

Catholicism, 85, 87, 153, 223, 302

Cawley, Harold, 91

Center for Constitutional Rights, 164

Central Intelligence Agency (CIA), 247

CFCs (chlorofluorocarbons), 135

Chad, 193

Chada, Raj, 154

Chauvin, Derek, 98

Chávez, César, 228

Chenoweth, Erica, 50–51, 119, 311, 316

Chevron, 231

Chichicastenango, Guatemala, 277

397

INDEX

INDEX

INDEX

INDEX

INDEX

INDEX

INDEX

INDEX

INDEX

Princeton University, 240

Pritchard, Amy, 207–8

property damage, 84–97, 151–2, 155–6

proportional representation, 125

proportionality argument, 151–7

Protect and Survive campaign (1974–80), 8–9

Protest and Survive (Thompson), 7, 219–20

'Protest—An Alternative View' (1988 lectures), 79–84, 88

Psych Declares, 117

psychological interventions, 194

public nuisance cases, 156–7, 206

Public Order Act (UK, 1986), 56, 59, 114, 145–51, 159

Public Order Act (UK, 2023), 164, 259, 326

public services, 303

Pugwash, 295

Quakers, 12, 59, 163, 204

Queen Elizabeth II Bridge, 199, 287, 321

Quisling, Vidkun, 227–8

racism, 98–104

anti-Semitism, 189–90, 202

Black Lives Matter (2013–), 3, 98, 159

Civil Rights Movement (1954–68), *see* Civil Rights Movement

slavery, 99–104, 188–9, 224–5, 226, 244–5

radical flank effect, 307–8

Radio Four, 56

Rainbow Warrior, 198

Ramblers' Association, 197

Randle, Michael, 58, 201–6, 229, 311

Rawls, John, 82–3, 151

Read, Rupert, 59–63, 64, 315

Reading, Berkshire, 142

Reagan, Ronald, 84, 186

Rebel Hive, 118

Rebel Verdict (Randle), 206

Reclaim These Streets, 155

Redmond, John, 89

Rees-Mogg, Jacob, 304

Reform Act (1832), 81

refugees, 34–5, 193–7, 291–3

Reid, Silas, 152, 156, 206–8, 312

Reith Lectures, 96

Renovate Switzerland, 323–4

Reykjavik Summit (1986), 295

Reznicek, Jessica, 87–8

Rhodes, Cecil John, 101

Richardson, Katherine, 321

roadblocks, *see* blockades

Rokeby Venus (Velázquez), 88

role playing, 13

Ronald Reagan, 295

Rose, Angus, 232

Rosebank, 322

Rowe, Christian, 158

Royal African Company, 99

Royal College of Defence Studies, 79, 159

Royal College of Psychiatrists, 117, 235

INDEX

INDEX

INDEX

INDEX